SCIENCE,
SCIENCE,
EVERYWHERE

SCIENCE, SCIENCE, EVERYWHERE

A Supplement to Childcraft—The How and Why Library

World Book, Inc.
a Scott Fetzer company
Chicago London Sydney Toronto

"Skyscrapers" from POEMS by Rachel Field (Macmillan, New York, 1957). Reprinted with the permission of Simon & Schuster Books for Young Readers, an imprint of Simon & Schuster Children's Publishing Division.

World Book, Inc.
525 West Monroe
Chicago, IL 60661

Library of Congress Cataloging-in-Publication Data
Science, science, everywhere.
       p.   cm.
    "A supplement to Childcraft—the how and why library."
    Includes index.
    Summary: Presents explanatory text and lab experiments through which to explore basic scientific concepts and topics at home, in forest and fields, in the air, in water, and in town.
    ISBN 0-7166-0697-6
    1. Science—Experiments—Juvenile literature.   2. Science—Juvenile literature.   [1. Science.   2. Science—Experiments.
3. Experiments.]   I. World Book, Inc.   II. Childcraft.
Q164.S2985   1997
500—dc21   97—3509

Printed in the United States of America

1 2 3 4 5 6 7 8 9 02 01 00 99 98 97

Contents

Staff

Preface

Do the words *science* and *scientist* make you think of glass tubes, microscopes, and white lab coats? Well, think again. You don't have to have fancy tools and special clothes to be a scientist. You can be a scientist anywhere—in your house, in your yard, and in your community—because there is **Science, Science, Everywhere!**

Science helps people answer questions such as, "How do soap bubbles get really big? Why does glue stick? How do animals communicate? Why doesn't a hot air balloon work with cold air? Where do rivers start and end?"

In this book, all kinds of scientists show you how they study the world. Science games, puzzles, riddles, and folklore add to the fun. Don't worry about not understanding unfamiliar words, they are defined in the margins and in the **Glossary**. The **Aha!** features and the **Breakthroughs!** highlight surprising science facts. When you are ready to try your skills, check out the **In Your Lab** sections. Turn to **Find Out More** to learn about other science resources.

Science is not just for professional astronomers, biologists, or chemists. Science helps all of us understand the world around us.

SCIENCE
ALL
AROUND

Science is everywhere—from the cereal you eat in the morning to the stars you see at night. Scientists are everywhere, too. Wherever there is something interesting and useful to find out about the world, there are scientists at work. Go outside. Observe how a river flows or how a helicopter flies and you just may find a scientist—you!

What Is Science?

The word *science* comes from a Latin word meaning "knowledge." Science is knowledge that results from our observation of the world around us. The knowledge must be tested again and again with experiments and proved true. This sets science apart from some other kinds of knowledge, for example, arts such as music, painting, and literature. There is no test that scientifically proves that one kind of painting is "more beautiful" or "more true" than another. No experiment can show whether a symphony is "right" or "wrong."

Science also is different from some other types of knowledge because in order to make progress, new scientific ideas often replace old ideas. For example, great works of art produced today do not take the place of masterpieces of the past. On the other hand, as new scientific discoveries are made, even many recent theories will be replaced by new ones. People once thought that Earth was flat, but through experimenting and observing, they proved that it was round.

Have you ever thought about what science means to you? Try to imagine your life without science. What would be missing? First of all, without science, you wouldn't know that Earth is a planet revolving around the sun. You would have no idea what causes lightning or thunder, either. And you never would have guessed that dinosaurs walked the Earth millions of years ago.

Science affects our daily lives in thousands of other ways, too. It provides the basis for much of our modern *technology*—the tools, materials, techniques, and power that make our lives easier. Without science, we wouldn't have televisions, radios, CD players, computers, airplanes, cars, telephones, or refrigerators. And modern medicines and surgical procedures would not exist.

Because there are so many things to observe and learn, there are many types of scientists. Think of some questions you could ask about the world around you. What makes bubbles form and float? What makes your desk lamp glow? These questions are part of a branch of science called *physics* (FIHZ ihks).

Although it is not possible to observe dinosaurs in action, this student is learning much about the prehistoric creatures by casting dinosaur bones.

Sometimes scientists need special tools to observe the world. These young scientists are using a microscope *(left)* to observe specimens and a telescope *(below)* to get a closer look at the moon.

Are you curious about the earth? Do you wonder about earthquakes and volcanoes? Then you have an interest in the science of *geology* (jee AHL uh jee).

People who study *astronomy* (uh STRAHN uh mee) explore the heavens—the planets, their moons, the stars, the galaxies, and other things in space.

Doctors study *anatomy* (uh NAT uh mee)—the various parts of living things and how they work together.

Chemistry and mathematics are two of the many more areas of science to be explored. Keep reading. Inside this book you'll meet all kinds of scientists who study interesting and exciting things.

Get an Attitude—
a Scientific Attitude!

No matter what area of science they study, most scientists have one thing in common—a scientific attitude. A scientific attitude is a way of thinking, observing, and testing things that helps lead to new understandings and discoveries. A scientific attitude is a mixture of curiosity, the need to know, an open mind, persistence (puhr SIHS tuhns), and carefulness. Do you have these qualities? Read on to find out more about them.

Curiosity

When you are curious, you're full of questions. Have you ever asked why something does what it does? Maybe you've wondered why a radio sounds better in some places than in others. Maybe you tried moving the radio from room to room and putting it in high and low places. If so, you are curious.

Stephen Hawking, a British scientist, is especially curious. His best-known work began with a deep curiosity about *black holes*, those dark bodies in space from which not even light can escape. Hawking's curiosity led him to make important discoveries about gravity and the universe.

Stephen Hawking

The need to know

In science, the need to know often begins with the desire to solve a problem. Have you ever grown a plant that withered or raised a tropical fish that got sick? To correct the problem, you probably tried to find out as much as you could about plants or fish.

In the same way, scientists search for ways to solve problems they care about. George Washington Carver wanted to help poor farmers in the Southern United States. Most crops wear out the soil after several years, so that the land is no longer good for farming. Carver experimented with peanuts and soybeans, two crops that improve the soil. He found many new uses for them. His work made it possible for farmers to plant and sell crops that would also restore their soil.

An open mind

Solving problems takes an open mind. Have you ever tried a new food or idea and it turned out to be surprisingly good?

Scientific knowledge often comes in surprising ways. American scientist Barbara McClintock studied patterns of color on different kinds of corn. She knew that kernel color and other *traits* (trayts) are passed along, just as hair color is passed along from parents to children. But McClintock discovered something surprising about the parts inside the cell that carry traits. These parts seemed to be able to suddenly change positions in the cell, creating new traits. Keeping an open mind, she conducted experiments for years to prove or disprove her new ideas. Her hard work paid off, and in 1983 she won the Nobel Prize for proving this new idea.

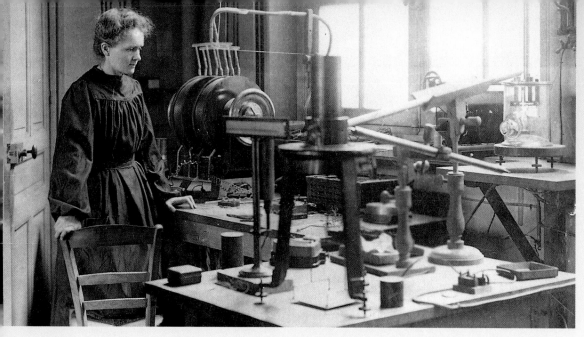

Marie Curie in her
laboratory

Persistence

Persistence means standing firm—never giving up. Suppose you try to find a planet with a telescope. You search the skies for many nights, but don't find anything. You keep trying. You finally see the planet. That's persistence.

Marie Curie had persistence. This famous French scientist won the Nobel Prize—twice! She discovered *radium* (RAY dee uhm), a radioactive metal. Radium is naturally found as a tiny part of a substance called *pitchblende* (PIHCH blehnd). For years, Curie worked to separate radium from pitchblende. In an unheated, leaky shed, she boiled pitchblende in huge pots. It took her four years and several thousand pounds of pitchblende just to get 0.004 ounce (0.1 gram) of pure radium.

Carefulness

Good science is careful science. Suppose you wonder if baking powder and baking soda cause the same reaction with vinegar. To find out, you might put each in a glass of vinegar. You'd be careful to use clean vinegar, clean glasses, and the same amount of each powder. You would also time and carefully observe each test.

English zoologist Jane Goodall became known for her careful, thorough work with chimpanzees. In northwestern Tanzania, Africa, she worked to win their trust through daily contact with them. Then she observed them at close range and wrote detailed reports of their behavior. Goodall discovered that chimpanzees make and use tools more than any other animal except human beings.

Now you know the formula for a scientific attitude. It's a mixture of curiosity, the need to know, an open mind, persistence, and carefulness.

Jane Goodall in her laboratory, the wild

Pack a Science Kit

If you have a hobby, you know how important special tools and supplies can be. Science is no different. You can probably find most of the things you need for your lab work around the house. You might want to keep the tools you use the most in a science kit. Just be sure to ask permission before you pack them!

One of a scientist's most important tools is a notebook or journal. In your notebook, you can write down your questions, your ideas about how to find the answers, and your theories about what will happen or what the answers will be. Your notebook should also include your observations and the results of your

July 30

First I filled the large bowl with water. Next I filled the small jar with water and 8 drops of food coloring.

experiments. It's important for a scientist to record detailed information so that it can later be shared with others. Your journal is a place you can make detailed sketches and write down measurements and notes. The more details you include, the better.

Other things that help you record what you learn are drawing paper, index cards, and colored pencils, pens, or markers. A grease pencil is good for writing on glass or plastic, and a camera lets you keep a record with photographs.

Here are some other types of things you may need.

Accurate measurements are very important. To measure things, you may use a medicine dropper, measuring cups and spoons, a ruler or tape measure, a stopwatch, or a thermometer.

Things aren't always in plain sight. To observe things, you may use a magnifying glass, flashlight, or binoculars.

To carry or observe specimens, you need containers of different shapes and sizes. To hold things, you may use jars and cardboard boxes, drinking glasses, disposable cups, bowls, a pitcher, or tweezers for holding small objects.

Some experiments require special equipment, such as safety goggles, a calculator, or small amounts of cooking ingredients, such as salt, flour, sugar, vinegar, cornstarch, and oil. You may also need paper towels, soap, cotton balls, aluminum foil, and drinking straws.

FLOUR

To study plants and animals, you may need potting soil, small clay pots, seeds, gardening tools such as a trowel and hand rake, a fish bowl or terrarium, a dip net, or field guides.

Many experiments involve attaching one thing to another. To attach things, you can use paper clips, tape, rubber bands, glue, string, or clay.

FIELD GUIDE

Every chapter in this book has science labs to help you start exploring different scientific ideas. Read an experiment carefully and make sure you have what you'll need before you begin. Each lab tells you what you need to get started, but you can substitute some items.

So what are you waiting for? Grab a backpack, a box, or a bucket and start putting together your own science kit!

Get into the Lab

A lab, or laboratory, is a place where scientists do experiments. It can be indoors or outdoors. An experiment is a test that a scientist does to check out an idea. When investigating their ideas, most scientists use six basic steps. They observe, ask questions, think of possible answers, test answers, draw conclusions, and retest. To learn about these steps, read on.

Observe and ask questions

Where do scientific ideas come from? They often begin with observation and curiosity. When you observe something and ask "Why?" or "How?" you have begun exploring scientifically. Suppose you notice that the deep water in your community's swimming pool is always colder than water near the surface. You ask why.

Hypothesize—think of possible answers

Scientists want to find answers to their questions. First, they *hypothesize* (hy PAHTH uh syz), or try to think of possible answers. In this case, you may think perhaps the cold water is pumped in down deep. Or, maybe warm water weighs less than cold water, so it rises to the top.

Choose one idea to test

Which explanation seems most likely? Sometimes scientists use what they already know to select the best idea. Sometimes they just have a hunch. You may find out that cold water is not pumped in at the bottom. So, you might decide to test the hypothesis that warm water weighs less than cold water.

Start experimenting

Scientists create experiments to test their hypotheses. Their experiments have variables (VAIR ee uh buhlz) and controls. *Variables* are the things that the scientist changes from one experiment to the next to see if a change makes the results different. A control is an experiment that has no changes or variables. Scientists may repeat experiments to test different variables. They compare the results to the results of the control experiment.

Suppose you invent an experiment to test your hypothesis. You may decide to dye some cold water red and to use some plain warm water. You pour the warm water into a large glass bowl and then quickly but gently lower in a small bottle of cold water. Your control experiment could be using plain and colored water that are exactly the same temperature. That means your variable in the other experiment is the temperature.

What do the experiments show you?

After scientists run an experiment many times, they may see patterns in their results. Then they try to form conclusions.

Suppose the cold red water in your experiment always settles to the bottom and the clear warm water always stays at the top. You might conclude: "Warm water weighs less than cold water, so it rises to the top and the cold water sinks." But if this doesn't always happen, you won't be able to reach a clear conclusion yet. Could the red dye make the cold water weigh more than the warm water? How can you find out?

What next?

If you cannot draw a clear conclusion, you may have to rethink your experiments. Perhaps you need to design them more carefully. For example, did you always put the cold water into the warm water at exactly the same speed? Or perhaps you'll use a different set of experiments or select a different hypothesis to test.

Each lab in this book is set up to help you develop your own

"Hey! I'll try an experiment!"

"What would happen if..."

ways of investigating and exploring ideas in science. It starts with questions relating to the preceding story. It gives a list of equipment and steps for performing an experiment. The section "Try it again and again" suggests variables to try. "Set up a control" suggests one way to produce a controlled result. "Now, write it down" helps you keep a record of your observations, conclusions, and questions.

Remember, it's your lab. It's up to you to decide what to test and explore.

Time to Think

"The time has come," the Walrus said,
"To talk of many things:
Of shoes—and ships—and sealing wax—
Of cabbages—and kings—
And why the sea is boiling hot—
And whether pigs have wings. . . ."

From *Through the Looking-Glass* by Lewis Carroll

Think Safety!

While science is fun, it's not exactly play. So it's time to talk safety. When you are doing experiments, always remember the rule, "Safety first." Help prevent accidents in the following ways:

- Before you begin a lab, read all instructions carefully. If you don't fully understand an instruction, ask a grown-up for help.
- Avoid spills, but just in case, line your work area with old newspapers before starting an experiment.
- While doing an experiment, do not eat or drink anything. Never, never put substances in your mouth.
- Keep first-aid supplies handy. Make sure you know where they are.
- If in doubt, don't do it. Taking chances can harm you or others.
- Ask a grown-up for help whenever you use chemicals, heat, sharp objects, or anything else that could hurt you.
- When you work with chemicals or flame, always wear safety goggles to protect your eyes.
- Don't fool around—it can be dangerous! Have fun, but treat your science experiments seriously.

Science Fun

Grab a friend and test your science knowledge.

You'll need

2 buckets or boxes
8 science tools, such as a notebook, a pencil, tape, a flashlight, a ruler, a watch, a measuring spoon, a field guide.

To play

1. Put the tools in a pile. Decide who goes first. Player #1 picks one of the numbered questions from this page or the next and reads it to player #2.

2. Player #2 picks one of the lettered responses from page 35. If it is correct, player #2 picks a tool out of the pile and puts it in his or her kit. (Answers are at the bottom of this page.)

3. Next, player #2 picks a question and reads it to player #1, and the game continues.

4. After all of the questions have been asked, the player who has the science kit with the most tools wins.

1. What is science?

2. What are six basic steps that are used in scientific investigation?

3. What are four science safety rules?

34

4. Why is science important?

Responses

A. An experiment that has no variables. Its results are compared with the results of experiments with variables.

B. Knowledge that results from observation; it can be tested with experiments and proved to be true.

C. Before beginning a lab, read all instructions carefully. Keep first-aid supplies handy. Don't fool around. If in doubt, don't do it. See page 33 for other possible answers.

D. Curiosity + the need to know + an open mind + persistence + carefulness.

E. There are many possible answers. Here is one: It is everywhere and affects everyone every day; it is the basis for much of our modern technology, such as televisions, computers, cars, and medicine.

F. Observe; ask questions; think of possible answers; test answers; draw conclusions; test and retest.

G. The French scientist who discovered the radioactive metal radium; she won two Nobel Prizes.

H. The things that change when an experiment is repeated.

5. What is a control?

6. What are variables?

7. Who was Marie Curie?

8. What is the formula for a scientific attitude?

FIELD GUI

SCIENCE
AT
HOME

Your home is full of science. From the orange juice in the fridge to the bubbles in the tub, science is at work. In this chapter you'll learn whether your juice is a solution, a colloid, or a suspension. You will find out why bubbles form better in soapy water than in plain water. You'll see how adhesives such as glue and paste do their sticky work. And you will read an *absorbing* tale.

Finding Solutions

Chemistry is the study of substances and the ways they can be changed or combined.

Sometimes it seemed that Connor was all elbows. On a visit to his Uncle Dan's house, one of those elbows knocked over a glass of milk. Uncle Dan taught chemistry in high school, and he took this opportunity to make one of his nutty chemist jokes.

"That's all right, Connor," said Uncle Dan. "No use crying over a spilled colloid (KAHL oyd) of water, sugar, and fat."

"A colloid?" Connor asked.

"Right," said Uncle Dan. "Many things you eat and drink aren't just plain liquids like water. Some, like your milk, are colloids. A *colloid* is a special kind of mixture. In a colloid, very tiny particles float around in a liquid or in another substance. In your milk, for example, tiny particles of fat, protein, and sugar are floating in water."

"But who cares about colloids, Uncle Dan?" asked Connor.

"Well, for one, you do. Let me show you why," explained Uncle Dan. He grabbed a jar and filled it with water and a handful of sand from a cactus pot in the window. "See how, after I stop shaking this mixture of water and sand, the sand settles at the bottom? That's what milk would be like if its particles of fat, protein, and milk sugar didn't form a colloid with water. I think I prefer my milk to be a smooth and creamy colloid—how about you?"

"Yes, I think I do like colloids after all," said Connor, "but what would you call that mixture of sand and water?"

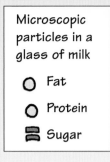

Microscopic particles in a glass of milk

O Fat
O Protein
▤ Sugar

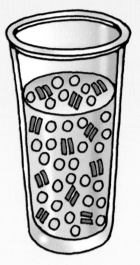

Marshmallows, whipped cream, and milk,
A pudding that's smoother than silk.
Would you be less than overjoyed
To know that toothpaste is also just a
C __ __ __ __ __ __ ?

answer on page 212

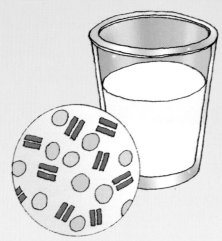

Colloid: milk

Suspension: sand and water

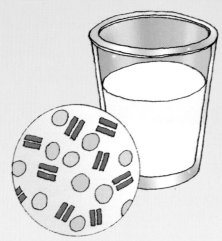

Solution: grape drink

"Good question," replied Uncle Dan as he got a carton of orange juice from the refrigerator and began to shake it. "I'm shaking the carton because orange particles have settled to the bottom, just like the sand settled in the bottle of water. Orange juice and the sand mixture are both suspensions (suh SPEHN shuhnz). In a *suspension,* particles of one substance float in another substance. But the particles are more dense than the substance they float in, so they settle to the bottom over time. And they can be *filtered* (separated) fairly easily from the substance they are floating in." He poured out a glass of juice.

For more information on water and filtering, see chapter 5.

"Now," said Uncle Dan. "May I pour you a refreshing solution (suh LOO shuhn) of water, sugar, and artificial flavorings?"

"Well, I don't know. . . ," mumbled Connor.

"You'll love this—it's Gulpy Grape." Uncle Dan poured out a glass. "A *solution* is similar to a colloid, but the particles in a solution are *dissolved* (dih ZAHLVD). That means they are broken up into super-tiny pieces about the size of molecules (MAHL uh kyoolz) and spread out evenly in the substance."

"I think I'm following you," said Connor. "But what's a molecule?"

"Sorry!" replied Uncle Dan. "A *molecule* is the smallest particle that a substance can be broken down into without changing it chemically."

"Now, let's continue with solutions. When you add sugar to your lemonade, you're actually making a solution. The sugar dissolves in the lemonade. I do the same when I add sugar to my coffee. The sugar dissolves in the coffee."

"In a colloid," he continued, "the small particles keep from settling because they float in a substance. But in a solution, the particles don't just float, they dissolve and spread into the substance."

"You can see the difference. Look at a glass of milk, a colloid. You can't see through it like you can see through Gulpy Grape, a solution. That's because the particles in a colloid, though tiny, are big enough for light rays to

*In a solution, the substance that's dissolved is called a **solute**. The substance that's doing the dissolving is the **solvent**. Sugar is a solute when it is dissolved in iced tea, the solvent.*

I am what I am when I'm shaken,
But I'm different when calm, no mistakin'.
A solution's not what you should mention.
When you name me, call me a

S___ ___ ___ ___ ___ ___ ___ ___ ___.

answer on page 212

What is smoke? Smoke is a colloid! It consists mainly of small particles of soot that remain after something is burned. These solids float along with other particles in a gas, the air.

bounce off them. When light rays bounce off the particles in a substance, the rays get mixed up so that you cannot clearly see through the substance."

"What are some other colloids, suspensions, and solutions?" Connor asked, sipping his juice.

"I'm glad you asked," said Uncle Dan. Then he started collecting things from around the house and setting them on the kitchen table.

What Is It?

Can you tell which is a colloid, a suspension, or a solution?

Item	Description
1. gelatin dessert	a solid (animal protein) dispersed in a liquid (water)
2. vinegar	two liquids (water and acetic acid) mixed
3. packing "peanuts"	gas (air) dispersed in a solid (plastic)

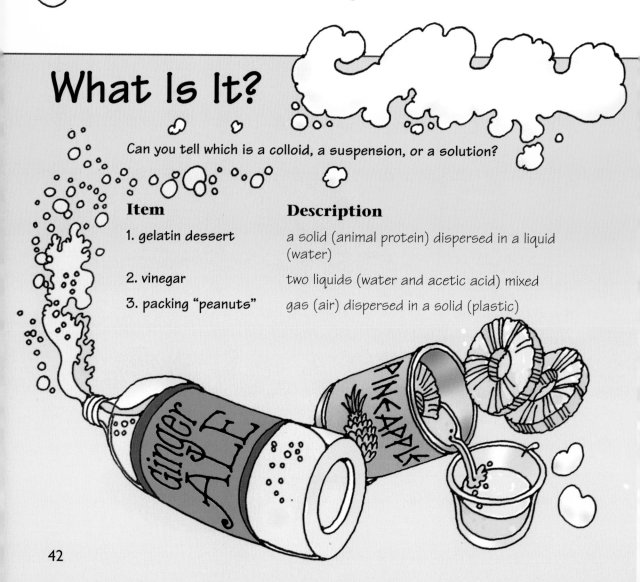

"All these things are either solutions, colloids, or suspensions. They are various combinations of solids, liquids, and gases," said Uncle Dan. "Let's make a list."

So Connor got paper and pencil, and he and Uncle Dan drew up the chart shown below. It describes some familiar colloids, solutions, and suspensions. Do you know what each item is? The answers are listed below.

Item	Description
4. ginger ale	solid (sugar), liquid (flavorings), and gas (carbon dioxide) dissolved in a liquid (water)
5. pineapple juice	solids (bits of pineapple) suspended in a liquid (water)
6. shaving cream	gas (air) dispersed in liquids (water, oils, and alcohol)
7. dirty water	solids (dirt particles) suspended in a liquid (water)

1—colloid, 2—solution, 3—colloid, 4—solution, 5—suspension, 6—colloid, 7—suspension

Mixing Things Up

Hmmmm . . . I wonder . . . Which substances make a solution with water, and which ones make a suspension? How can I tell the difference between a solution and a suspension?

GATHER TOOLS

- 5 large jars the same size and 5 drinking glasses
- warm water
- 5 spoons the same size
- a small can (6 fl. oz.) of orange juice concentrate (without pulp)
- flour
- salt
- food coloring
- 5 coffee filters, all the same kind
- 5 rubber bands
- a flashlight

Set up and give it a try

1 Fill the jars halfway with water. Put a spoonful of salt into jar #1 and stir thoroughly. Stir a spoonful of flour into jar #2. Stir several spoonfuls of the orange juice concentrate into jar #3. Squeeze a few drops of the food coloring into jar #4.

Be sure to use the same amount and temperature of water in each jar. Stir the mixtures equally.

2 Shine the flashlight into each jar. Which jars can you clearly see through? In which ones can you see particles?

44

3 Pair an empty glass with each jar. Put a filter in each glass. Use a rubber band to hold the filter in place.

4 Carefully pour one-third of the mixture in each jar through the filter into its paired glass. What happens in each glass?

Set up a control

Set up another jar and glass and use plain water.

5 Leave the jars with their remaining mixtures undisturbed overnight. What do you see in each jar the next day?

Try it again and again

- Mix other materials with water—for example, sand, cornstarch, or sugar.
- Replace the water with vinegar or oil.

Now, write it down

Write or draw what you saw in each jar. Which jar held a solution? a suspension? How do you know? Did anything you tried make a colloid? What other items can you test?

Compare your notes with the ones on page 212.

Bubble Science

Connor had enjoyed finding out about solutions, colloids, and suspensions with Uncle Dan. Now Uncle Dan stood at the kitchen sink putting detergent in the dishwater. Connor watched it foam and grow into mounds of sparkling suds.

"The suds remind me a little of shaving cream," said Connor. "Are suds a colloid, too?"

"Not exactly, but that's a good guess," said Uncle Dan. "Suds are really just a lot of bubbles stuck together. But we can learn a lot from bubbles."

"Such as what causes them?" asked Connor.

"That's a start," said Uncle Dan. "To understand bubbles, we need to understand a property of water called *surface tension*. You see, water is made up of molecules. These water molecules are attracted to one another almost like iron to a magnet. And, because all the molecules are pulling on one another, the water stays together. This sticking together is called *cohesion* (koh HEE zhuhn)."

The water's surface tension keeps this pond skater from sinking.

"Because of cohesion, the molecules at the surface of the water are attracted to the water molecules below and beside them rather than to the air above them. As a result, a sort of skin forms on the top of the water."

"A skin?"

"Well, it's not like your own skin. But it's a surface that takes some force to break. Have you ever noticed insects skating around on a pond?"

"Yes, I saw one at Crescent Lake last summer," replied Connor.

Uncle Dan continued. "Those insects are called water striders. They actually use the surface tension to walk on water."

"Cool," said Connor.

"Wait," said Uncle Dan, "I'll demonstrate surface tension." He got a small glass out of the cupboard and pulled a handful of pennies from his pocket. He filled the glass with water nearly all the way to the top and handed Connor the pennies. "Here, Connor. Carefully drop the pennies, one at a time, into the glass of water."

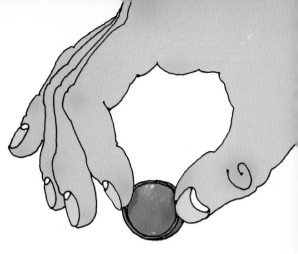

Connor dropped one, two, three, and more pennies in the glass. The water rose above the rim in a bulge. "Oh, I see," he said. "Surface tension keeps the water from spilling over the sides."

"Right," said Uncle Dan. "Surface tension and adhesion (ad HEE zhuhn). *Adhesion* is what holds water molecules to the rim of the glass."

I am skin without pimples
Or freckles or dimples.
Do I need to mention
That I'm S __ __ __ __ __ __
T __ __ __ __ __ __ ?

answer on page 212

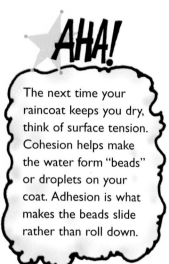

AHA!

The next time your raincoat keeps you dry, think of surface tension. Cohesion helps make the water form "beads" or droplets on your coat. Adhesion is what makes the beads slide rather than roll down.

"So where do bubbles come in?" Connor asked.

"Let's figure it out together," said Uncle Dan. "First, what shape are water drops?"

"Kind of round like a ball, I guess," answered Connor.

"Right. And what shape are bubbles?"

"Round, too," answered Connor.

"Right again. Surface tension makes all the water molecules pull toward each other. If no container is forcing the water into a shape, a small amount of water naturally takes a round shape." Uncle Dan ran water into another glass. "See how air bubbles form in the water? But when I turn off the water, most of the bubbles disappear. That's because surface tension pulls the bubbles apart and lets the air escape."

"But the sink with the detergent in it is still full of bubbles," said Connor.

Water molecule

Detergent molecule

"You're one step ahead of me, as usual," said Uncle Dan. "Detergent has a special property. A molecule of detergent has one end that is attracted to water molecules and one end that is repelled by them. So detergent molecules wedge themselves between water molecules and push the water molecules apart. This is how detergent cleans your dishes. One end of the detergent molecule clings to dirt and the other to the water that washes the dirt away."

"This special property of detergent also weakens water surface tension and allows the surface of water to stretch more easily. The stretching allows bubbles to form better and grow bigger. Detergent also keeps water from evaporating. That's what makes soapy bubbles last longer."

"So a soap bubble is just a skin of detergent and water, with air inside," said Connor.

"Exactly!" Uncle Dan beamed. "How big do you think bubbles can grow? Let's find out."

Evaporation *is the process by which liquid turns into vapor. Wipe a shiny surface like a cool kettle with a damp cloth. Watch the water disappear. The water is now present in the air in the form of water vapor.*

Breakthroughs!

Some of the world's most famous scientists dabbled in bubble research. However, perhaps the world's greatest bubble researcher was Joseph A. Plateau of Belgium.

Plateau, born in 1801, recklessly threw himself into scientific research when he was a young man. To test an idea about human vision, he stared into the sun for up to 25 seconds at a time—something that is extremely harmful. As a result, Plateau gradually lost his eyesight. He became totally blind by his early forties.

1820's

1800

1873

Despite this setback, Plateau completed his most important work in his later years with the help of friends and family members who helped him study the way soap bubbles join. They examined thousands of groups of bubbles, measuring the angles their surfaces formed where they met. Plateau discovered something fascinating: bubbles connect at one of two angles, either 109° or 120°. In 1873, he published a book on these findings, *Statique des Liquides*.

Joseph A. Plateau

Plateau's discovery led many mathematicians to study bubblelike surfaces. They wanted to prove that because of surface tension, water molecules cover as little area as necessary. A bubble has the same quality. During the 1930's, a team of American mathematicians produced the first mathematical proof that this is true.

1930's

1995

Americans Joel Hass and Roger Schlafly made another important bubble discovery. They found that a container shaped like a double bubble takes up less space than any other container used to hold two equal volumes of different liquids. The double-bubble shape could reduce the area needed for fuel tanks in rockets. Try making the shape with your bubbles at home!

NOW

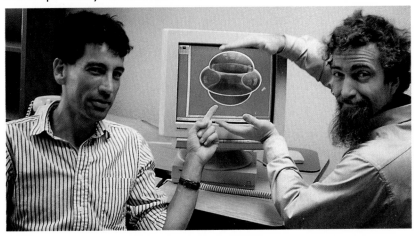

Joel Hass and Roger Schlafly

Breaking Tension

> Hmmmm . . . I wonder . . .
> How can I see surface tension in action?
> How strong is it?

GATHER TOOLS

- 6 small bowls of water
- 2 needles
- 2 paper clips
- 2 pieces of aluminum foil, cut in 1-inch (2.5-cm) squares
- small pieces of facial tissue or toilet paper
- liquid soap
- toothpicks

Set up and give it a try

1 Carefully wash, rinse, and dry all of the bowls, needles, and paper clips.

Rub a needle between your fingers to make its surface a little oily. Place a piece of tissue *carefully* on the surface of the water in one of the bowls. Right away, drop the needle *gently* on the tissue. The paper will soak up the water and sink, but the needle will float.

2 Use more tissue to *gently* float the paper clip on the water surface in another bowl. Float the foil square by itself in the last bowl.

Set up a control

Set up three more bowls with the same materials. Use the clean ends of the toothpicks.

3 Dip one end of each toothpick in a drop of liquid soap. Dip the soapy end of a toothpick into each bowl. What happens?

Try it again and again

- Use different amounts of soap.
- Use different kinds of soap—for example, chips from a bar of soap or grains of laundry detergent.

Now, write it down

Jot down your observations. Note what you did to each bowl of water and the item in it. How strong would you say surface tension is? What does it take to weaken surface tension?

Compare your notes with those on page 212.

Sticky
Situations

Uncle Dan and Connor were sitting at the kitchen table when the door opened and in walked Aunt Judith. She sat down at the table holding her umbrella.

"Uncle Dan has been teaching me all kinds of interesting things about bubbles and colloids and stuff. What happened to your umbrella?" Connor asked.

"I slammed the handle in the car door," Aunt Judith replied. "I hope I can repair it."

"Let's try the superglue," suggested Uncle Dan, opening the drawer. "Hey, why don't you tell Connor about adhesives, Judith?"

"Yeah!" said Connor. Aunt Judith was a scientist, too. She worked for a company that made plastics.

"Well, Connor, *adhesives* are substances, such as glue, paste, and tape, that we use to stick one thing to another," said Aunt Judith. "They work because of two related forces—*cohesion* and *adhesion*."

"I know what those are," Connor said. "Cohesion is the force that holds a substance together, and adhesion is the force that holds a substance to something else."

"Very good!" said Aunt Judith. "And what makes this glue adhesive so useful is that it starts out as a liquid. A liquid has less cohesion than a solid, so we can spread it on things in a nice thin layer. But more important, when an adhesive is in liquid form, its molecules can get very close to a surface. In fact, these molecules *bond* (attach) to the molecules of the surface. Now, what's the name of the force that makes things want to stick to other things?"

AHA!

Why doesn't superglue stick to its own container? Because there's no moisture inside the tube! Superglue needs moisture to stick. Your skin has moisture on it, so it's a perfect surface for bonding with superglue. Be careful when you use this glue—don't get stuck on yourself!

Glue is an adhesive. It can bond things together because of adhesion and cohesion.

"Adhesion!" cried Connor.

"Right again," said Aunt Judith. She opened the tube of superglue and picked up her umbrella. "As I spread the superglue, it begins to bond with my umbrella handle. Then, when I attach the piece that broke off, the adhesive forms a bond with that surface. When the adhesive dries, its own molecules link together, forming a hard film with very strong cohesion. Then my umbrella will be fixed!"

"Fixed forever?" Connor asked.

"It should be. Some adhesives are so strong that it's easier to break the thing you've repaired than to break the adhesive bond."

"Are all adhesives that strong?" asked Connor.

"Not all of them," replied Aunt Judith. She picked up a pad of sticky notes. "The adhesive on these is designed to be just a little bit sticky. The notes stick only for a short time. And you can easily pull them off."

"Pretty neat," said Connor.

"Oh, and here's another kind of adhesive." She waved a roll of stamps. "Stamps and envelopes have a special kind of adhesive called *gum*. Unlike superglue, you can make gum sticky again and again. After it dries, just moisten it. Then you can peel it and restick it. Gum forms a bond strong enough to fasten paper. If you've ever steamed open an envelope

I'm sticky when you lick me,
I dry out when you stick me.
I help your letters come.
You can't chew me up,
but I'm G __ __

answer on page 212

and then sealed it back up, you've seen gum in action!"

"We made paste from flour and water at school once," said Connor.

"Yes," said Aunt Judith. "Flour is a type of starch. Mixing starch with water makes a sticky paste. Starches are good adhesives because their molecules can attach to other things."

For information on how stretchy steel building materials are, see chapter 6.

"Our teacher once told us that glue was made from boiled animals."

"Yes, that's true. We use the word *glue* for all kinds of adhesives. But only adhesives made from animal parts are really glue," replied Aunt

Judith. "When skins and bones are boiled, they release a protein called gelatin. It's the same stuff that gelatin desserts are made of. Gelatin is naturally sticky. Glue makers dry the gelatin, crush it into powder, and add other ingredients. You have to add water before you use it. Other adhesives are made from artificial products such as nylon, which is a type of plastic."

"Look at all of the things around here that use adhesives," called Uncle Dan with his arms full.

Soon, Connor and Aunt Judith found themselves in a sticky situation as they explored the pile of tape, bows, bandages, note paper, stamps, and other sticky things.

AHA!

The rubber tree is the source of a natural adhesive called *latex* (LAY tehks). The tree has its own use for this substance. When a hungry insect chomps on a rubber tree leaf, it gets a sticky mouthful. Unable to chew, the insect has its last meal!

Your Own Super-Adhesive

> Hmmmm . . . I wonder . . .
> Can I make my own super-adhesive?
> Can I make it from common
> substances we have around the
> house?

GATHER TOOLS

- disposable cups
- measuring cup
- measuring spoons
- teaspoon
- coffee filters
- milk
- white vinegar
- baking soda
- rubber bands
- test materials such as jars, pencils, paper clips, toothpicks, rubber bands, and papers

Set up and give it a try

1 Pour 1/2 cup (118 ml) milk into one of the disposable cups. Add 1 tablespoon (14.5 ml) of vinegar to the milk and stir. For best results, use clean utensils.

2 Place a filter in a cup. Use a rubber band to hold it in place. Pour the milk mixture into the filter. Let it drain for half an hour, or until the solids are as dry as you can get them. (Drain twice if necessary.)

3 Throw away the liquid that runs out of the milk mixture. Then put the solids in the cup. Rub the solids between your fingers. Would they make a good adhesive?

5 Test this glue for adhesion on your test materials. (Get permission to use the items.) How well does it work? On paper, is it as strong as the glue on an envelope or more like a sticky note?

4 Add a teaspoon (5 ml) of baking soda to the remaining solids. Stir well. Cover the mixture and let it sit for 24 hours.

6 Is the glue waterproof?

Try it again and again

• Use heavy cream instead of milk and lemon juice instead of vinegar.
• Try replacing the baking soda with other substances, such as detergent, milk of magnesia, or crushed-up chewable antacid tablets (ask for permission first).

Set up a control

If you want to know whether more of one ingredient—such as vinegar or milk—makes your glue stronger, divide a batch into three clumps. Add a little more vinegar or milk to one clump, a lot more to another clump, and none to the third. Compare the glues.

Now, write it down

Jot down the exact recipes you used and describe the type of glue each one made. Note which recipes worked best on which substances—for example, on paper, wood, plastic. Which ingredients made the best glue? Why do you think that is?

Compare your notes with those on page 212.

An Absorbing Tale

Connor had a question for Aunt Judith the next time he visited her.

"Aunt Judith, last night Dad gave Mom some carnations—you know, those frilly flowers. Some of them were green. I had never seen a green flower, but Mom said they were dyed. When I looked at them, I could see that the dye was inside the flower! How can this be?"

"The answer is related to something you've already learned about," replied Aunt Judith. "Remember the word *adhesion?*"

"It's the way things stick to other things," said Connor.

"Yes, and cohesion is the way things hold together themselves," continued Aunt Judith. "Well, adhesion and cohesion can be pretty powerful forces."

"That's really interesting. But I still don't understand how my mom's carnations turned green," said Connor.

"I'm getting to that," replied Aunt Judith. "A plant's stem has tiny tubes called *capillaries* (KAP uh lehr eez). When water enters a capillary, its molecules are more attracted to the walls than to each other. This attraction, called *capillarity,* lifts the water upward against gravity. Capillarity is how water climbs from roots deep underground to the tops of the tallest trees."

Water travels up a plant stem through these tiny tubes called capillaries.

White carnations dyed blue

For more information on how plants get water, as well as sunlight, see chapter 3.

"Now I think I understand how florists turn carnations green," said Connor. "They mix green dye with water and let capillarity carry dye up the stems to the petals."

"Wow!" said Aunt Judith. "I'm going to tell your mom how smart you are."

"I think she already knows," said Connor, shyly.

"Capillarity is very useful," said Aunt Judith. She got a paper towel and a glass of water. "Between its many fibers, this paper towel has millions of tiny spaces that are like tubes. Water is absorbed into these spaces in the towel by capillarity. Watch! When I put the end of the paper towel in the water, the water rises up into it. "

"The bath towels soak up water," observed Connor. "They must use capillarity, too."

In a tube tinier than a hair,
Water climbs up without a stair.
An amazing feat in all clarity,
I'm talking about
C __ __ __ __ __ __ __ __ __ __.

answer on page 212

"Yes, and so do facial tissues and some kinds of clothing. When we need to take up moisture, we use these absorbent materials. Capillarity lifts water up a tube until it can't lift the weight of any more water. In a larger tube, the water cannot climb as high because the larger amount of water weighs much more. That is why capillarity occurs in small tubes,

AHA!

Why are cotton clothes comfortable to wear on hot summer days? Cotton fibers are long, hollow tubes. On a hot day, the cotton fibers in your shirt and shorts quickly draw sweat away from your skin, keeping you comfortable.

Which are more absorbent, the tan colored, unusually shaped natural sponges, or the artificial sponges that are colorful and rectangular? Test for yourself the next time you wash the car or take a bath.

where the weight of the water is smaller, and the water can be lifted higher."

"Do other things have capillaries?" asked Connor.

"For centuries," Aunt Judith continued, "people have used sea animals called sponges for bathing and cleaning. Sponges have soft skeletons made of an absorbent material called *spongin* (SPUHNJ ihn). Spongin is full of tiny tubelike spaces."

"Do you mean that the sponges we use to clean our car were sea creatures?" asked Connor.

"Your sponges are probably not natural ones," said Aunt Judith. "Most of the sponges we use today are made from artificial materials. They're made in factories to have properties similar to natural sponges."

Just then Uncle Dan walked in with a frown on his face. "What's wrong?" asked Connor.

"Well, when I sat down to have lunch in the park, I put my coat over the edge of the bench. I didn't notice it at the time, but the end of my coat must have been in a puddle. Now look! Most of my coat is soaking wet."

"Hey," exclaimed Connor, "I bet your coat is all wet because capillarity was in action. The water climbed up spaces in your coat just like water climbs up capillaries in paper towels."

"My thoughts exactly," said Aunt Judith.

Capillarity

Hmmmm . . . I wonder . . .
How fast can molecules climb
through capillaries?
How much work can they really do?

GATHER TOOLS

- paper towels
- facial tissues
- paper napkins
- toilet paper
- 4 small, clear glasses or jars
- food coloring
- a spoon
- water

Set up and give it a try

1. Fill each glass one-fourth with water.

2. Add a few drops of food coloring to each glass and stir.

3. Fold each absorbent material into 4 layers. Put a folded material in each glass. Which do you think is most absorbent? least absorbent?

4 After 1 minute, observe the strips and the height of the water on each paper. Were your predictions correct?

Try it again and again

- Compare blotter paper, coffee filters, and tissue paper.
- Use rubbing alcohol or nail polish remover instead of water. (Get permission first.)
- Put the stem of a white carnation in colored water for a couple of days.

RUBBING ALCOHOL

NAIL POLISH REMOVER

MINUTES

Now, write it down

What did you observe? How fast did the water move up each of the papers? Which material was the most absorbent? How does it differ from the other materials in the way it looks and feels?

Compare your notes with those on page 212.

SCIENCE
IN
FIELDS
AND
FORESTS

When you picture scientists, do you imagine them working in big laboratories, surrounded by specialized equipment? Well, lots of scientists do, but many other scientists work outdoors in fields and forests. If you think like a scientist, you too can learn a lot from the plants, animals, and other living things outside your door. Just take the time to look, listen, and smell.

How Do Plants GROW?

You and your friends eat fruits, vegetables, and other foods to grow up strong and healthy. But how do fruits, vegetables, and other plants grow? I happen to know because I'm a botanist. Plants use sunlight and simple ingredients such as water to make the food they need. However, plants, can't cook, of course. When they make their own food, the process is called *photosynthesis* (FOH tuh SIHN thuh sihs). In photosynthesis, a plant uses the sun's energy to convert water and a gas called *carbon dioxide* (KAHR buhn dy AHK syd) into

A **botanist** (BAHT uh nihst) is a scientist who studies plant life.

sugar. Imagine not having to turn on an oven, ever! Well, it's not quite *that* easy.

Plants work constantly to gather what they need to make food. In fact, plants can actually move toward the things they need to make it. For example, vines in forests get to sunlight by climbing up trees. They cling to the tree trunks and then grow upward to reach the light. These movements that plants make are called tropisms (TROH pih zuhmz).

AHA!

Stroke the tip of a pea vine or a cucumber vine with a stick and watch the tip turn. You're seeing thigmotropism in action! *Thigmotropism* is the tendency of vines and some other plants to wrap around a support.

A *tropism* is a plant's response to certain signals in the world around it. The movement of a plant toward light is called *phototropism*. Scientists call the plant's response to gravity *geotropism*. And different parts of a plant react in different ways to the pull of gravity. For example, the root of a plant responds to gravity by growing downward. This movement sends the root into the soil, where it can absorb water and minerals. The plant's stem also responds to gravity, but it reacts by growing upward instead of downward. This movement pushes the stem above the soil, where leaves can spread and absorb the sun's rays.

Have you ever seen a potted plant tipped on its side by accident? You may have noticed an upward bend in the plant's stem. Because of

geotropism, the stem started growing straight upward again—away from the pull of gravity. This made the stem crooked! You could not see the root, but it bent, too. Because of geotropism, it grew downward.

How did the root and the stem sense that they had been tilted off course? Like all living things, plants are made up of *cells*—tiny bits of living matter. Plants have special cells that sense gravity. They are found along a plant's stem and in the tip of its root. Each of these cells contains "sacks" of grains that sink to its lowest part.

As long as the plant is straight, the sacks lie at the bottom of the gravity-sensing cells. But when the plant is tilted, the sacks shift. Scientists think the shifting sacks start an uneven flow of chemicals that affect the plant's growth.

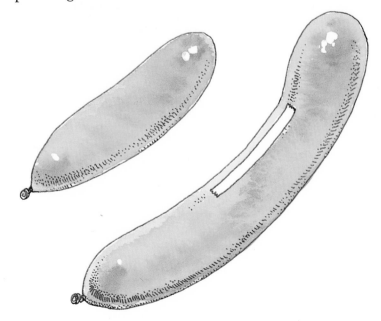

Blow up a balloon partway, then stick a piece of tape onto it. Continue blowing up the balloon and you will see that the tape lets one side expand, or grow, more than the other side. Just as the tape affects the way in which the balloon expands, chemicals in plants affect the direction in which roots and stems grow.

A chemical that promotes growth, called *auxin* (AWK sihn), flows to the lower side of the stem. This makes the stem's lower side grow faster than its upper side. As a result, the stem bends and grows upward. In the root, a chemical that *slows* growth flows to the lower side. It slows the lower side's growth, but the upper side keeps growing. As a result, the root bends downward.

Gravity isn't the only signal that plants respond to. Green plants also respond to light and water, which they need to make food.

For example, a plant on a sunny window sill may bend toward the light. That's because when one side of a plant receives more light than the other, auxin starts flowing to the dark side of the stem. The dark side grows faster than the sunlit side, and the plant bends toward the light. However, the same plant is still also affected by the pull of gravity. While it bends toward the sun, the stem is also growing upward. So the plant grows at an angle.

Although roots do not respond to light, they do grow toward water as well as gravity. Their movement toward water is known as *hydrotropism*. Hydrotropism sometimes pulls more strongly on roots than gravity. In very dry conditions, a root may grow sideways—toward an underground river, for instance. The roots of many desert trees spread out close to the surface to capture every drop of rainfall they can.

Plants don't look busy. They are so quiet, and they seem so still. But now you know that plants are constantly on the move, responding to signals that help them get the things they need to make their own food.

Seeking Light

Hmmmm . . . I wonder . . .
How strongly does a bean
plant seek light?
What if it doesn't get
any light?

Set up and give it a try

1 Soak the beans overnight in water. Fill a pot with soil and plant a few beans. Keep the soil moist. Wait 5 to 7 days for a sprout.

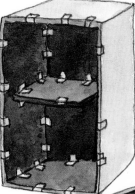

2 Line the inside of the box with black paper. Cut the paper a little large. Then tape the paper in the box. Fold in the extra paper to help keep out light.

3 Cut an index card so that it is 1 1/2 inches (4 cm) shorter than the width of the box. Cover the card with black paper and tape it inside the box like a shelf. Cut a 1 1/2-inch hole near the top end of the box top.

GATHER TOOLS

- runner beans or dry pinto beans
- planting pots or cups
- potting soil
- empty shoe boxes or cracker boxes
- scissors
- black construction paper
- index cards as wide as the box
- tape

4 Place the pot in the box under the shelf. Close the box and tape it shut.

5 Place the box on a sunny window sill, with the hole facing the sun. Each day, open the box, observe the plant, and water it if it is dry. Then close the box.

Set up a control

Grow a bean plant in an identical box without a hole to see how it grows without light. Grow a bean plant on the same window sill without a box.

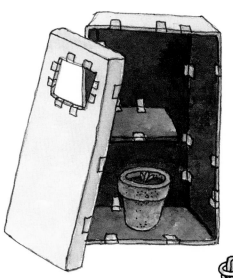

Try it again and again

- Use lamplight instead of sunlight. (Ask an adult for help to set it up safely.)

- Add another shelf.

- Use various kinds of plants.

Now, write it down

Keep notes on what happens daily for two weeks. Include when and how much you water your plants, how much sunlight is available, and how the plants look. Can you now answer the questions listed at the beginning of this lab? Do you have other questions? Keep testing.

Compare your notes with those on page 212.

Plant Poisons
Can Be Helpful

H. M. Lock here. I'm a biochemist on a mission. My job is to find useful substances in plants. This may surprise you, but some of the plants that help us the most are poisonous.

Scientists classify thousands of plants around the world as poisonous. A poisonous plant is one that contains a substance that can harm people or animals if they touch or swallow it. Even some house plants can be dangerous. For instance, the leaves and stems of the philodendron (fihl uh DEHN druhn) have crystals that burn your mouth and throat. Plant poisons can also produce problems such as skin rashes, stomachaches, and liver damage.

*A **biochemist** (by oh KEHM ihst) is a scientist who studies the chemicals and chemical processes in living things.*

Poisons are the plants' defense against plant-eaters. The poisons make these plants smell or taste bad, so many animals avoid them, including humans. After all, plants can't run away when in danger as animals do.

These defenses do not always work, however. Some animals don't mind the smell and taste of poisonous plants, and some of these animals aren't injured by plants that make people sick. For example, some rabbits produce a chemical in their bodies that changes certain plant poisons into harmless substances. Also, animals with four chambers in their stomach, such as deer, can eat bushels of poisonous plants with no ill effects.

Yuck!

Even people can handle low doses of some plant poisons. Cooks and doctors use small amounts of these chemicals in foods and medicines. For example, cinnamon,

Peppermint

Nutmeg

Cinnamon

peppermint, nutmeg, and many other flavorings from plants contain substances called *volatile* (VAHL uh tuhl) oils. In high doses, volatile oils can make a person feel dizzy, nauseated, or panicky. But of course, people don't eat large amounts of these volatile oils because the flavorings are quite strong.

One group of plant poisons is known as *alkaloids* (AL kuh loydz). They are found in larkspur, glory lilies, belladonna, and many other garden plants. Eating these plants can cause numbness, vomiting, and even death. Yet doctors use alkaloids from glory lilies to treat serious diseases such as leprosy and cancer. And eye doctors drop an alkaloid from belladonna

The science called **pharmacognosy** *(FAHR muh KAHG nuh see) deals with the chemical substances in natural drugs. Some of these drugs come from poisons in plants.*

Some poisons are found only in certain parts of a plant. The tuber of the potato—the part you bake, mash, or fry—is perfectly safe to eat, and so are the stems of rhubarb in rhubarb pie. But rhubarb leaves, and all the green parts of the potato plant, are poisonous.

Larkspur

Belladonna

into patients' eyes to enlarge their pupils for eye examinations. *Ipecac* (IHP uh kak) syrup, another medicine, contains alkaloids from a plant in the coffee family. Ipecac causes vomiting. A dose of ipecac syrup is sometimes given to make a person vomit a poison.

Another group of plant poisons can make a person's skin very sensitive to sunlight. These poisons are *phototoxins* (FOH toh TAHKS ihnz). Just touching a plant that has phototoxins and then going out in the sun can cause redness, itchiness, and blisters. But doctors use some kinds of phototoxins to treat a skin disease called psoriasis (suh RY uh sihs). Psoriasis patients take pills containing phototoxins and then expose their skin to ultraviolet light—like the light in the sun's rays. This treatment often heals the flaking skin caused by psoriasis.

To be "plant smart," follow these tips:

✔ Learn which plants in your house and neighborhood are poisonous.

✔ Eat only what you have permission to eat. Remember, some parts of a plant may be poisonous even though you can eat other parts.

✔ Keep babies and pets away from house plants.

✔ Stay away from burning plants. Breathing in their smoke may be bad for you. Smoke from burning poison ivy, for example, is especially dangerous.

✔ When cooking over a campfire, don't roast hot dogs or other foods on branches from poisonous plants or plants you don't know are safe.

✔ Make sure your family's first-aid kit has instructions for treating a poison victim.

✔ Know the phone number of your local poison control center or put the number by your home telephones.

An Outdoor Habitat

An **ecologist** (ee KAHL uh jihst) is a scientist who studies the relationships between living things and their habitats.

How well do you know your nearest neighbors, the plants and animals in the yard, garden, or park near your house? I ask because I'm an ecologist. My name is Barry E. Cology, and I study plants and animals everywhere I can, even in my own backyard. My backyard is a habitat for many living things. A *habitat* is the area of land, air, or water—or all these places—where a living thing dwells.

You say you want to know more about *your* neighbors? Well, I'll share my science secrets so you can see how ecologists study habitats.

First, an ecologist needs to mark off a small area of the habitat to be explored. Just as nobody can examine all of Earth at once, I can't closely examine all of my backyard at once. So, I set up a transect and a quadrat (KWAH druht). It's easier than it sounds.

To make a transect, I drive a *stake,* or pointed stick, into the ground. Next, I walk 10 feet (3 m) and put in another stake. Then I connect the stakes with string. This straight string above the ground is my *transect*. Be careful not to trip!

Next, I make a quadrat, which can look like a pie or a checkerboard. For example, a square *quadrat* is a 3-foot (about 1-m) square made of wood, screws, string, and tacks that form 16 smaller squares. I toss the quadrat under the transect. Now I have marked off an area of my yard that I can easily explore.

When I study my neighbors, I always bring along a backpack with graph paper, colored pencils, a magnifying glass, and *field guides* (illustrated books about wildlife). On the graph paper, I draw straight lines to represent the transect. Each square stands for some number of inches or centimeters on the transect.

AHA!

In the United States, it's illegal to kill or capture endangered butterfly species, such as the Schaus swallowtail and the San Bruno elfin. International laws make capturing butterflies illegal in other countries, too. So now, butterfly enthusiasts and butterfly specialists, *lepidopterists (lehp uh DAHP tuhr ihsts),* plant flowers to attract butterflies. Then they can observe and photograph the lovely creatures up close, without having to capture or kill them.

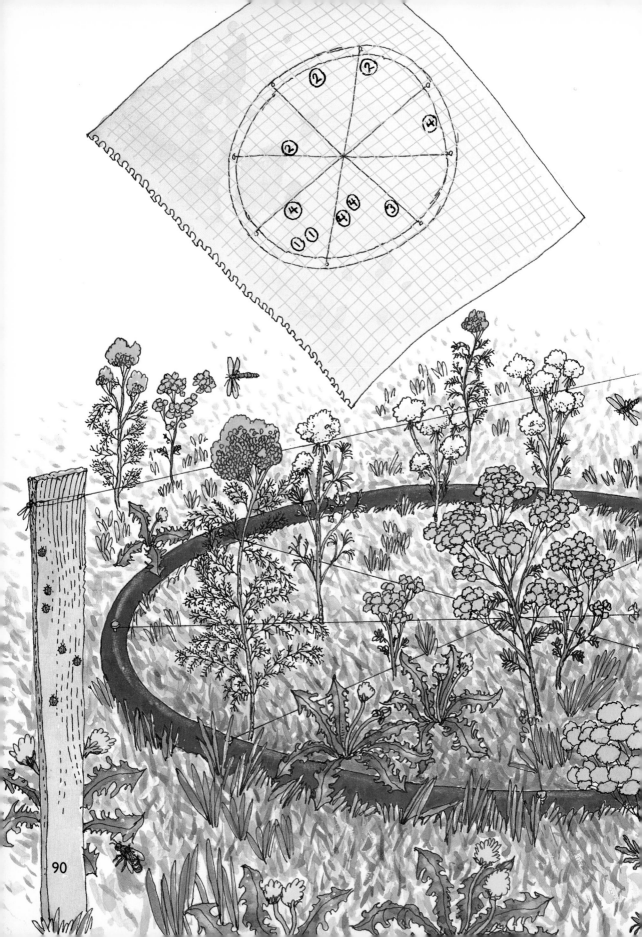

Field Guide

Dandelion

Queen Anne's lace

Tansy

Try your skill at identifying wildflowers. Can you match Barry's wildflower sketches to photos in this field guide?
answers on page 212

AHA!

It's not surprising that insects and plants that live at ground level experience a warmer climate than we do. After all, the earth absorbs heat from the sun all day and only *s l o w l y* releases it.

I sketch plants that lie under the transect. Then I sketch the insects that live on the plants. On another sheet of graph paper, I draw a map of the quadrat.

On my maps, I label the plants and animals I know. To identify unfamiliar ones, I look for clues. Take wildflowers, for example. Their leaves and petals are usually arranged in set patterns. When I find a wildflower, I note its color. I carefully sketch its leaf and petal patterns. Then I compare my sketch with pictures in a field guide. Usually I find a match.

I write down more than just names, though. I also record the *population* (number) of each kind of living thing within the quadrat. And I describe differences in the soil. Later, I compare this information to other areas in my yard.

The more information I collect, the more I can find out about the plants and animals that live in that habitat. Humans usually feel climate conditions 3 feet or more above the ground. But plants and animals that live at ground level may experience a different climate, and animals in the trees can experience yet other climates. These climates that exist in small parts of a habitat are called *microclimates*. For example, rocks shade and cool parts of the ground, and trees block winds.

For more information on weather, see chapter 4.

To study microclimates, I make a temperature pole. I start with an old broomstick whittled to a point. I drive the point into the ground along the transect. Then I tie three cardboard tubes to the broomstick— one at ground level, one a foot (30 cm) high, and one at 3 feet. Three thermometers are tied so that one falls inside each tube. The tubes shield the thermometers from the sun. I check the thermometers several times a day and record the temperatures on my transect map.

Finally, I lie on the ground and just watch the wildlife along the transect and in the quadrats. I try to look at the habitat through the eyes of a small creature. I imagine being a butterfly. Fluttering in the air, I watch bees sucking nectar and look for a flower for myself. I spot a beetle nibbling petals, aphids sipping sap, and a wasp chewing pollen. The bee lands on a blossom and prepares to sip nectar. Suddenly, a hungry yellow crab spider sidles into view and the bee is off again.

Web-Weaver Habitats

Hmmmm . . . I wonder . . . Where do spiders in my backyard habitat weave their webs? Are all webs the same? Are the locations of different webs alike in any way?

Caution! Don't touch or play with any spiders that you find. Most spiders do not harm humans, but some spider bites are painful, and a few are even dangerous.

GATHER TOOLS

- colored pencils
- drawing paper
- a hard book or clipboard to put paper on when writing

IMPORTANT: Get an adult's permission before doing this activity.

Set up and give it a try

1 Visit your backyard or a park (ask permission first). Draw a diagram of the area on a sheet of plain paper. Include trees, bushes, walls, tables, and play equipment. Then search for spider webs. Check under slides, steps, and ladders, as well as in low and high corners of supporting bars and beams.

2 Each time you find a web, mark its location on your diagram by writing a number (start with #1 and go up). Copy that number on a separate sheet of paper. Next to the number, sketch the web. Write a description of the web's location using words such as *shady, sunny, dry, damp, hidden,* or *exposed.*

Tangled web

Orb web

3 Identify each web's type by comparing it with the web types shown here.

Funnel web

Sheet web

Set up a control
Search for webs in other yards or parks.

Try it again and again

- Be sure to go web searching at a site both on a sunny day and then again after a rain or storm.

- Map your basement, attic, or garage and locate webs there.

Now, write it down

Compare your descriptions of the webs you found. Are they all the same? Are they all found in the same type of location? What can you conclude about spider habitats?

Compare your notes with those on page 213.

Life at the Hive

Hey, you there! Keep your distance. Sorry, I didn't mean to scare you. But if you want to get any closer, you'll need a suit like mine. I'm Melliferous, the beekeeper. I'm an entomologist (ehn tuh MAHL uh jihst). And I can tell you it's not good to bother honey bees when they are working.

Usually, if you let bees alone, they'll let you alone, too. But when one honey bee is threatened, hundreds swarm to its defense. Honey bees work together and depend on one another. Each member of their colony, or community, does an important job.

How have I learned all this? Well, I've learned a lot from little Bea, a worker in this colony. You'll recognize her by the

Worker

blue dot I painted on her back. We entomologists often mark insects we want to study with painted dots or stickers. I watched Bea develop from a tiny egg laid by the queen bee.

The *queen* bee is unlike any other bee in the colony. She is the only bee in the colony that lays eggs. In fact, that is the queen bee's only job. In the spring she lays up to 2,000 eggs a day. She deposits each one in its own six-sided wax cell in the hive. The queen fertilizes some of the eggs with fluid she receives from a male bee. The eggs she fertilizes become *workers* (females). The others become *drones* (males). Our friend Bea was one of the fertilized eggs.

Queen

Three days later, little Bea hatched as a larva.

Drone

Worker bees called "nurses" made a creamy food called *royal jelly* and fed it to Bea and the other larvae. Royal jelly is very rich in nutrients. In a few days, Bea's diet changed to honey and pollen.

For eight days, Bea grew inside her cell. Then she stopped growing, and worker bees sealed her inside her cell. She became a pupa and stayed in

Larva *is the name of an insect when it has just hatched and begun growing. The plural of larva is* **larvae** *(LAHR vee).*

Pupa *(PYOO puh) is the name of the stage when the insect stops growing and changes into an adult.*

her cell while her body changed. Two weeks later, Bea emerged as a full-grown worker.

I could tell that Bea was a female because she had a stinger. A drone emerged from a nearby cell. He looked large but, like all male honey bees, he had no stinger. He didn't need one, because the hundreds of drones in a hive never work. They do not even feed themselves. Their sole purpose is to mate with a queen from a different hive.

After three days as an adult, Bea began working as a nurse. Other workers her age became house bees. These bees keep the hive tidy. They also clear wax out of the cells so they can be used again.

Soon Bea graduated to wax bee. Now she was busy building new cells in the hive and repairing old ones. She produced wax from wax glands in her body. Then she chewed the wax pieces and molded them into cells and combs.

Honey bees use cells to store honey and pollen as well as to raise their young. Wax bees build the cells tilting up. This prevents the honey and pollen from running out.

Bea's next job was guard bee. She kept a sharp lookout for wasps who feed honey bees to their young. If an enemy had tried to enter the hive, she would have stung it and died.

After three weeks as an adult, Bea became a *forager* (FAWR ihj er), or food gatherer. The foragers are the busiest and most important workers in the colony. They buzz about from flower to flower, collecting pollen and nectar.

On Bea's first outing, she found a patch of sweet purple clover 150 feet (46 m) from the hive. She used

Combs are sheets of wax made up of the honey bees' six-sided cells. *Honeycombs* are used to store the hive's honey.

AHA!

Worker honey bees live several weeks to several months, or until they sting an enemy. When the bee stings, hooks on the end of the stinger hold tight in the enemy. The stinger pulls out of the bee's body, and the bee soon dies. Queen bees have smooth stingers, so they don't die when they sting. They can live up to five years.

As a honey bee collects nectar, it becomes covered with pollen.

AHA!

Everyone huddle! When cold weather arrives, the colony gathers around the stored honey. The bees in the center feed while those in the outer circle keep the eaters warm. Then the bees change places, and the outside bees eat while the former "in crowd" keeps them warm.

her tubelike tongue to suck the clover's nectar— its sweet juice. She kept the nectar in her honey stomach, which is inside her body just in front of her true stomach. Pollen clung to the feathery hairs on Bea's front legs. Her two middle legs brushed the pollen off the hairs and into pouches, or pollen baskets, on her hind legs.

For information on water insects, see chapter 5.

Back at the hive, house bees took Bea's pollen and nectar. They rolled the pollen into wads of beebread and packed the beebread in storage cells.

Bea poured her nectar into the house bees' mouths. They mixed the nectar with their saliva to make honey and stored it in cells. Then they fanned the honey with their wings to thicken it. Finally, when the honey was just right, they sealed the storage cells with wax.

Bea did a "round dance" for the rest of the foragers. Her fast movements described the rich food source close to the hive. Excited workers headed out to find the clover, and Bea paused just long enough for a beebread snack. Then she took off for another "honey run."

To be a bee or not to be a bee—
What kind of bee would I want to be?
Perhaps a cleaning house-bee bee
With a little bee apron and broom?
Or a guard bee with a sharp bee stinger
To guard the honey room?
Perhaps I'd search for crimson clover
To make the honey and sweet beebread—
Once, again, and over and over—
And then to sleep in a little bee bed!

Breakthroughs!

Honey bee dances may be the closest thing in the animal world to human language. Austrian scientist Karl von Frisch discovered this marvelous behavior in 1949 and studied it for many years. At first, most other scientists refused to believe that insects could communicate intelligently.

Martin Lindauer, a co-worker of von Frisch, published *Communication Among Social Bees* in 1961. In this book, Lindauer explained that honey bees also use dances to tell a colony of new locations for a hive.

1961

1940

1949

Karl von Frisch discovered that honey bee foragers perform dances that tell the other bees where to find nectar. A "round dance" tells of food close to the hive. A "tail-wagging" dance tells of a distant food source. This complex dance includes a figure eight with a wiggle in the middle. The speed, direction, turns, and wiggles tell in which direction— and how far—the bees must fly to find the food.

Tail-wagging dance

Round dance

Thomas Collett

English scientist Thomas Collett discovered that bees use memories of landmarks, such as trees and ponds, as well as magnetic material that works somewhat like a compass in their stomach to find their way back to places. Scientists hailed these findings as important in understanding insects.

1994

NOW

1973

Karl von Frisch shared the Nobel Prize for his studies of animal behavior. In addition to honey bee dances, he studied vision in fish.

Today

Scientists continue to study honey bees. An American biologist, Dr. Scott Camazine, has tagged more than 4,000 honey bees with tiny numbers. He plans to track each bee, hoping to learn more about where a bee goes, what it does, when it does it, and how it communicates with other bees.

Karl von Frisch

103

What's That Squirrel Saying?

A **zoologist** *(zoh AHL uh jihst) is a scientist who studies all aspects of animal life.*

Pheromone *(FEHR uh mohn) is a chemical substance released by many kinds of animals to communicate with other members of their species.*

Hello! Can you stay for a chat? I love to communicate. As a zoologist, I enjoy seeing how animals communicate. At first, I admit, their "conversations" didn't seem much like ours. But then I learned that animals communicate through sounds, movements, and scents called pheromones. I only had to figure out what these things meant!

Most animals don't think about what they want to communicate. Instead, they act largely by instinct. In other words, they behave in ways that they have inherited from their parents and grandparents. These ways help them survive. For example, a ground squirrel whistles when it sees an animal that hunts it, the hawk, circling above. The whistle is a medium-pitched sound, a difficult sound to locate. It doesn't reveal the squirrel's position. But other squirrels hear the warning, and they scramble to their burrows.

When a badger slinks nearby, the same ground squirrel chatters an alarm. The chatter is a mixture of low- and high-pitched sounds. These sounds are easier to locate than medium-pitched ones. As a result, other ground squirrels know where the badger is. They run and hide. As the badger passes each hiding place, another squirrel chatters, revealing the badger's new location. This

Instinct (IHN stingkt) *is something a person or animal does or knows without having to learn it. For example, babies don't have to learn to smile when they are happy. They do it because of instinct.*

A ground squirrel whistles to signal that a hawk is circling above.

behavior shows how animals cooperate to survive. The frustrated badger gives up and goes away hungry.

Prairie dogs, cousins of ground squirrels, are great communicators, too. When two prairie dogs meet, they sniff each other's faces. They identify relatives by their scents. If the prairie dogs are family members, they may "kiss" again. If their scents are unfamiliar, they chatter and chase each other around.

A deer makes relatively little noise, but it communicates in other ways. For example, when a *buck* (male deer) looks for a mate, he may paw the ground, urinate, and deposit a

An axis deer leaves his scent on branches to mark his territory.

A rabbit warns other rabbits of danger by thumping the ground.

Wolves communicate with facial expressions—just like you do. When you're happy, you smile. When a wolf is happy, its tongue hangs out loosely and its ears point forward. But if a wolf wrinkles its nose, shows its teeth, and perks its ears up straight—watch out! It's ready to fight.

pheromone, or scent, produced between his toes. This scent attracts nearby *does* (dohz) (female deer). A buck also leaves a scent when he rubs his forehead against tree branches. This scent warns other bucks to stay off his territory.

Rabbits release pheromones to mark their territory. When they sense danger, rabbits warn others by thumping the ground with their hind feet. Then they hop away fast, looking for cover.

Do you want to know what animals have to say? Then look, listen, and smell when they come near. Maybe you will learn something.

SCIENCE
IN THE
AIR

You can't see it, smell it, or taste it, but life on this planet would be impossible without it. It helps people ride bikes and helps helicopters fly. It can pump water and generate electricity. And if we know it well and watch it carefully, we can see how it affects our weather every day. What is this wonderful stuff? Air!

How Thin Is Thin Air?

Do you think air is a lightweight subject? Think again. You can study air all your life and still be amazed by it. In this chapter, you'll learn a lot about air, including why it's important to know so much about it.

Air is invisible, but that doesn't mean there is nothing to it. Air is a mixture of gases that swirls around our planet. Wave your hand through the air. Did you feel anything? You just waved your hand through gases, mostly nitrogen and a lot of oxygen. Air also has *water vapor* (water in the form of a gas), carbon dioxide, and other gases. These invisible gases weigh more than you might expect. They are like blankets that circle the earth, and gravity pulls on them. The weight of these blankets as they press down on the earth is called air pressure or *atmospheric pressure* (AT muh SFIR ihk PREHSH uhr).

At sea level, a low point on the earth, there is a lot of air pressure because there is a lot of air above it pressing down. At the top of a mountain, there is less air pressure simply because there is less air above it, and so less air pressing down.

Temperature also greatly affects air pressure. When air is cool, its gas molecules squeeze together tightly. Because cool air is dense, it is heavier than warm air. *Dense* means closely packed together. When air is warm, the gas molecules fly apart, making the air thin and light, less dense. So warm air weighs less than cool air, and as a result, puts less pressure on the earth. If you have ever seen people fly in a balloon, you've seen hot air at work. The thin, light hot air inside the balloon makes it float.

Air's weight also depends on its *humidity,* that is, on how much moisture it contains. You might think that the more water air holds, the heavier it

AHA!

would be. But the opposite is true. Air with a lot of moisture in it is less dense. Something that is less dense is packed together less closely. So air that is less dense is also less heavy. This means that moist air is lighter than dry air. The heaviest air is cool and dry. The lightest air is warm and moist.

Earth is covered with a constantly changing pattern of high-pressure areas and low-pressure areas. These are known as *pressure systems,* and they are responsible for our weather.

When the air pressure is low in a certain area, meteorologists call that area a "low." You've probably seen *LOW* or *L* on weather maps. The air around a low is usually warm and moist, so clouds and rain are likely.

An area where the air pressure is high, or heavy, is called a "high." Have you seen *HIGH* or *H* on weather maps? The air around a high is usually dry and often cool, so sunny skies and clear weather are in the forecast.

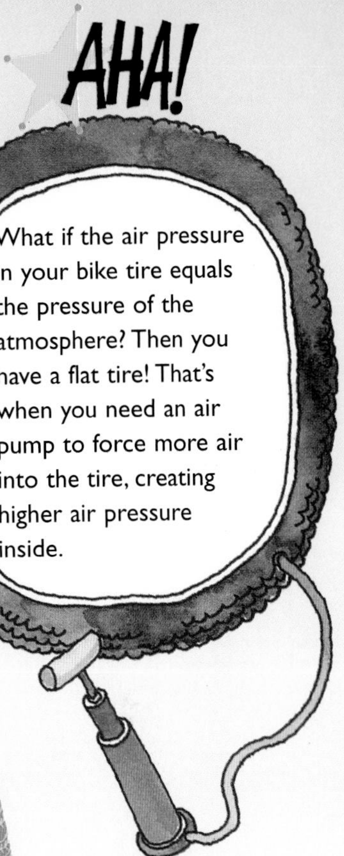

AHA!

What if the air pressure in your bike tire equals the pressure of the atmosphere? Then you have a flat tire! That's when you need an air pump to force more air into the tire, creating higher air pressure inside.

*Meteorology is the scientific study of Earth's atmosphere and weather. A **meteorologist** is a scientist who studies meteorology. These words come from* meteora, *a Greek word meaning "things in the sky."*

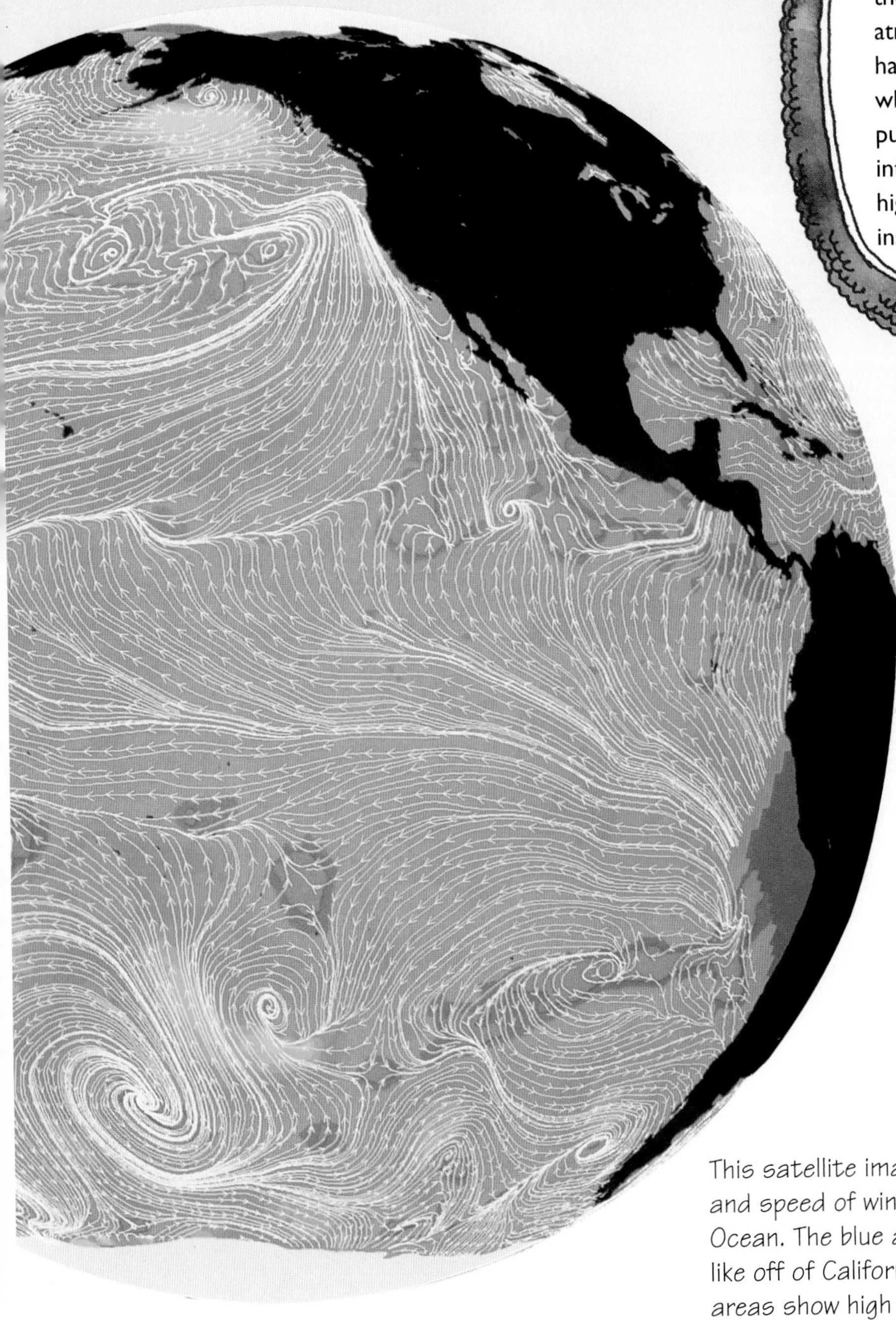

This satellite image shows the direction and speed of winds over the Pacific Ocean. The blue areas show slow winds, like off of California. The red and orange areas show high wind speeds, like the ones in the South Pacific.

113

Some people say they can detect changes in the air pressure according to the onset of a headache or aches in their joints. However, most scientists and other people use a *barometer* to measure air pressure.

There are two types of barometer—the mercury barometer and the aneroid (AN uh royd) barometer. In the mercury barometer, a liquid metal called mercury partly fills a hollow glass tube. The tube sits in a pool of mercury in a shallow dish that is open to the air. When air pressure is high, the heavy air presses down on the mercury in the dish,

Mercury barometer

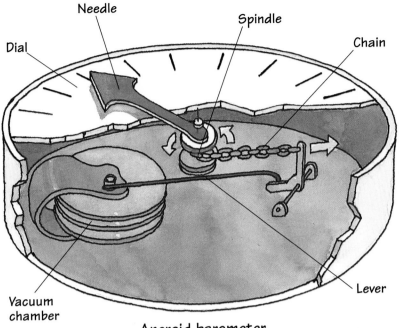

Aneroid barometer

forcing the mercury up the tube. When air pressure is low, the mercury in the tube drops and runs back into the dish.

The *aneroid barometer* contains a vacuum chamber—a metal container with no air inside. A drop in air pressure makes the chamber *expand* (get bigger) as the walls of the chamber move farther apart. A rise makes them *contract* (get smaller) as the walls are pushed closer together. The chamber is connected to a needle that moves around a dial, which we can read.

A high reading on a barometer indicates good weather. A low reading means that a storm may be on the way.

Today the barometer is staying nice and high. Looks like a great day for a picnic!

When you are standing up, the weight of about 2,000 pounds (907 kg) of air is pressing down on you! When you lie down, that weight is even greater. Luckily, you are supported by air pressure on all sides. Otherwise, you'd really feel a weight on your shoulders!

IN YOUR LAB
Make and Use a Barometer

Hmmmm . . . I wonder . . . The weight of the air pressing down on Earth is always changing. But how can I measure it? How much does altitude affect air pressure?

GATHER TOOLS

- rubber balloon
- 2 identical glass jars
- rubber bands
- a paper soda straw
- glue
- a flat wooden toothpick
- a pencil
- scissors
- a craft stick

Set up and give it a try

1 Cut off the narrow end of a balloon—the end with the blow-hole—and throw it away. Save the rest of the balloon. Stretch the balloon piece tightly over the mouth of a jar. Fasten it tightly with a rubber band.

2 Flatten the straw. Cut one end into a pointed tip. Glue the other end to the balloon piece at the center.

3 Glue the toothpick to the balloon right at the top edge of the jar so that the straw rests on the wood. The straw is your barometer's "needle."

4 Using rubber bands, attach the craft stick to the outside of the other jar so that it sticks about 1 inch (2.5 cm) above the top edge of the jar.

Set up a control
Use a barometer from the store as a control. Or, watch TV weather reports and record the barometric pressure for your area.

5 Place the jars close together so that the sharp end of the "needle" points to the craft stick on the second jar. To avoid getting false readings, make sure your barometer is away from heat, air conditioning, and open windows.

6 Check your barometer every day at the same time. Mark the spot on the stick where the needle points, and write the date beside it.

Try it again and again

• Take readings on the same day from high and low places, such as a hilltop or valley.

• Take readings for several weeks or for several days during different seasons.

Now, write it down

After a week, compare your barometric records with the ones from your control. Did your readings change when they did? Why do you think your readings match or don't match?

Compare your notes with the ones on page 213.

How Do We Use Air?

People cannot control the air outside and how it affects the weather. However, people can use the air in many helpful ways.

Air packs power. Prove this by blowing up a balloon. But don't tie the end in a knot—just let go of the balloon. As the air rushes out, watch the balloon fly across the room. The balloon is propelled by air power! We can use this power in different ways.

We use exhaust fans to force stale air out of a room. An exhaust fan pulls air out instead of blowing air in. Exhaust fans are used to ventilate rooms where air often becomes too hot, moist, or smelly. The kitchen or bathroom in your home may have an exhaust fan.

Air power can also be used to dry things. Try hanging wet clothes on a clothesline on a warm, windy day. The water in the wet clothes *evaporates* (turns into water vapor) in the warm,

dry air. The same thing happens when you dry
your hair with a hair dryer.

We use moving air to preserve foods, too.
Many of the *microorganisms* (living things too
small to be seen except with a microscope) that
spoil food cannot grow without moisture, so we
can preserve some foods by drying them. In
food-processing plants, trays of food pass
through long tunnels in which hot, dry air is
blowing. The food comes out dried—and
preserved.

Windmills are designed to catch the wind.
They use its power to do heavy work. The
spinning blades of windmills provide energy for
water pumps. Windmills also turn *generators*—
machines that produce electricity.

And how about a helicopter? You guessed it—helicopters work with air to overcome gravity. Picture a helicopter. It has one or more rotors spinning on top. A *rotor* is a set of blades that spins around.

The word **helicopter** comes from Greek words meaning "spiral" and "wing."

Each rotor blade is curved on the top and flat on the bottom. This shape causes air to rush faster over the top than against the bottom when the rotors are whirling. As a result, the air pressure becomes greater below the rotor than above it. This difference in air pressure creates a force called *lift*. Lift raises the helicopter into the air and keeps it flying. Some helicopters also have a small rotor on their tail. This rotor helps stabilize the 'copter— that is, keeps it from wobbling or jerking about.

To learn how a fish or a boat stays upright in water, see chapter 5.

Did you ever think we could use air in so many ways to make our lives easier, more comfortable, and more exciting? Air can do some pretty heavy work!

AHA!

People throughout history have had ideas for a flying machine like a helicopter. For example, around A.D. 320, a Chinese writer described a rotor-powered flying machine. This design may have been based on a Chinese toy called the flying top. Such toys used feather rotors to fly.

Seattle
49/38

L

Winnipeg
28/10

Montreal
43/36

H

Billings
32/20

Bismarck
29/12

L

Boise
40/27

New York City
58/42

Salt Lake City
42/32

Cheyenne
39/30

Chicago
49/32

H

Pittsburgh
59/45

Francisco
/44

Las Vegas
69/45

Albuquerque
64/35

St. Louis
52/34

Norfolk
65/43

es
0/53

Charlotte
69/43

North
Atlantic
Ocean

H

Birmingham
77/62

L

orth
cific
an

Houston
80/58

Miami
79/70

Gulf of Mexico

The *Front*
of a **Storm**

Hi! I'm weather forecaster Rayne Gauge
(GAYJ). It's my job to figure out what the
weather will do next. To do so, I keep a close eye
on air masses. Among other things, they help
me tell you where a bad storm is lurking, and if
today is a good day for a picnic.

You see, the air is not the same wherever you go. The world is covered by enormous clumps of air called *air masses*. An air mass forms over an area in which the temperature is fairly constant. The mass can cover hundreds of square miles and can be wet or dry, cold or hot, cool or warm. Dry, cold air masses develop over the dry, frigid poles. Wet, hot air masses develop over tropical oceans. Dry, hot air masses form over deserts.

Earth's rotation naturally causes air masses to move from one area to another.

Earth's rotation naturally causes air masses to move from one area to another. And as you already know, air masses take on the temperature of the area over which they move. However, they do so very slowly because of their great size. So before an area can greatly change an air mass, the air mass affects the area's weather.

Let me explain. When a cold air mass and a warm air mass meet, you have a zone that is called a *front*. Most changes in weather occur along fronts.

AHA!

A front may be the reason that you are wearing a bathing suit while someone a few hundred miles away needs a parka. A cold front often divides sharply different air masses. Sometimes the temperature difference can be 50 °F (28 °C) or more.

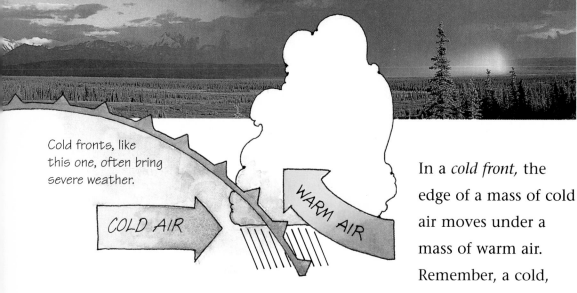

Cold fronts, like this one, often bring severe weather.

COLD AIR

WARM AIR

In a *cold front,* the edge of a mass of cold air moves under a mass of warm air. Remember, a cold, dry air mass weighs more and has higher pressure than a hot, wet one. If the two masses meet, the heavier cold air flows beneath the warm air, forcing the warm air to rise quickly. As the warm air rises, it cools. The water vapor in it may form rain or snow.

A *warm front* occurs when a warm air mass catches up with a cold air mass that is moving in the same direction. The warm air flows upward over the cold air, cooling as it rises. Its water vapor forms clouds

that may travel hundreds of miles ahead of the
front. These clouds slowly thicken, bringing
rain, or maybe sleet or snow.
And the wet weather may
hang on for a while.

To learn how
buildings stay
strong in stormy
weather, see
chapter 6.

The weather around a cold
front is usually more severe
than the weather around a warm front.
However, warm front weather will probably take
longer to clear out. Do you see a front coming
this way?

Warm fronts,
like this one,
usually take a
while to clear.

WARM AIR

COLD AIR

How Warm and Cold Air Behave

Hmmmm . . . I wonder . . .
Is it true that warm air expands
and rises?
How do I know that cold air really
contracts and sinks?

GATHER TOOLS

- identical soda or juice bottles
- identical balloons
- bowls or pans
- hot water
- cold water
- ice cubes

Set up and give it a try

1 Place one balloon over each bottle's mouth.

Set up a Control

Make another bottle and balloon setup, exactly like the other two, and put it in water that is room temperature.

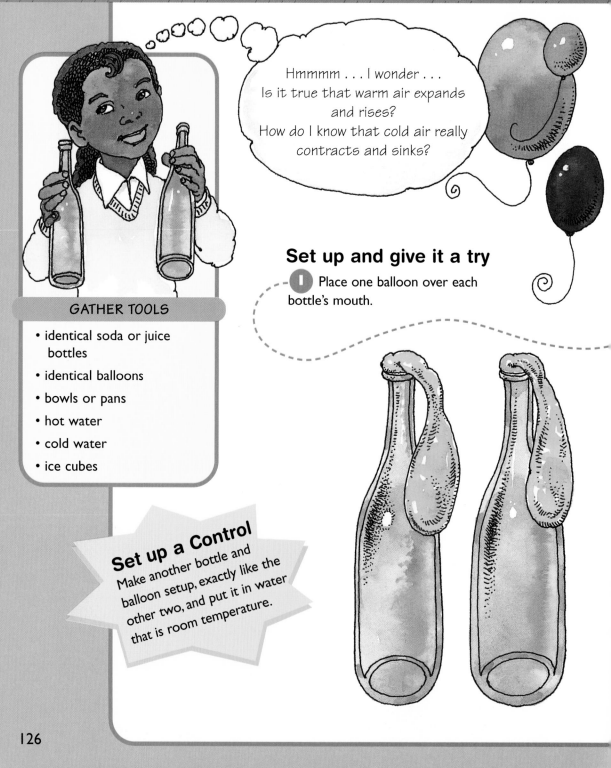

2 In one bowl pour about 2 inches (5 cm) of hot water from the tap. In the other bowl pour ice water to the same depth.

3 Place one bottle in each bowl and watch what happens!

Try it again and again

- Redo the experiment without the bottles; blow up and tie the balloons and put them in water and ice.

- Try using water of different temperatures. Use a thermometer.

Now, write it down

Jot down what happened to each balloon and why you think it happened.

Compare your notes with the ones on page 213.

Tracking Violent Weather

Howdy, Partner! Some folks think of me as a modern cowgirl, er, cowperson. Instead of chasing bucking broncos on the range, I track bouncing twisters. Basically, I'm a meteorologist who's interested in tornadoes.

You may have noticed that I called the whirling storm a *twister*. Most of the time people use the word *tornado*. Both words mean the same thing, but I like the picture that the word *twister* conjures up. As a matter of fact, it looks like we have twister weather today—a great time for a storm chase and a good time for me to share some basic information about tornadoes.

First, you should know that we weather forecasters watch for violent weather around the clock. And we have some pretty fancy equipment to help us. For instance, weather satellites in space send us incredible photographs of the cloud tops. Weather balloons swirl in the upper atmosphere. They carry equipment that collects data such as temperature and air pressure. Radar stations on the ground pick up images of precipitation and storms.

Precipitation *is any form of water that falls from clouds to the ground—rain, snow, sleet, or hail.*

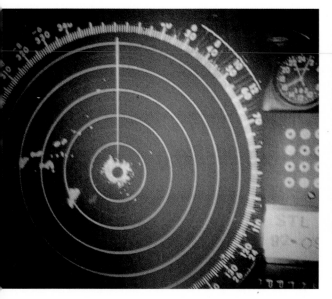 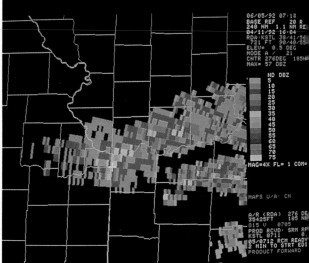

Perhaps the most exciting tool we have is *Doppler radar.* This special kind of radar works like the radar police use to catch speeders. It detects sudden changes in wind direction and speed. With Doppler radar we can see a tornado forming deep inside clouds and then warn people that severe weather is on the way.

Early radar equipment shows a storm's size and location (above left). But Doppler radar provides much more detail (above). The colors depict the speed and direction of winds, temperatures, and the types of precipitation. This storm produced golfball-sized hailstones!

AHA!

Tornadoes are powerful storms that can do a lot of damage—yet most tornadoes last less than an hour and travel an average distance of about 20 miles (36 km).

Speaking of severe weather, look what's been brewing while we've been talking about tornadoes. That line of thunderstorms looks really mean! Our Doppler radar tells us that one of these is a supercell thunderstorm. A *supercell* is a towering storm that lives longer than others. It has an *updraft* (rising air) that rotates. Sounds like the start of a twister, doesn't it?

Tornadoes form in the same kind of air that cooks up thunderstorms. Usually, the air is hot and *humid* (moist), which makes it unstable. That means it's likely to rise rapidly. And if a cold front approaches, the heavier cold air flows under the warm air mass and pushes it up. Rising air may turn into a column of twisting air called a *funnel cloud*. If the air column grows stronger and stronger, it may become a tornado!

Up ahead is a wall cloud. A *wall cloud* is a dark, rounded cloud at the base of a thunderstorm. No rain falls from it because the rising wind is too strong. A funnel could drop down at any time. If it does, then we have a tornado.

Not all funnel clouds "touch down," or drop to the ground. But when they do, they can do a lot of damage. Meteorologists measure a tornado's strength by the damage it leaves behind. We use the Fujita scale to rate tornadoes from F0 (the weakest) to F5 (the strongest).

Those of us who have been close to a twister can tell you that they are frightening. The updrafts near the funnel can pick up cars and trucks. The whirling dust is so thick that it's hard to see what's going on. Hail often falls, and sometimes the hailstones are as big as eggs!

AHA!

Why is a tornado's strength measured by the damage it leaves behind? That is largely because it's difficult to predict where the middle of the storm will be, and it's dangerous to get equipment there.

Tornadoes are the most powerful storms on Earth. That's why it's important to watch out for them. They can pack winds of more than 200 miles per hour (320 kph). Most tornadoes are skinny, several hundred yards or meters in diameter. But some monsters get to be 1 1/2 miles (2.4 km) wide.

Take my advice. Don't run after tornadoes—or any other violent storms. Storm chasing is a job for professionals who understand the risks involved in studying violent weather. We hope we learn enough to keep everyone safe from twisters.

Do you know where tornadoes are most likely to strike? The central United States encounters more tornadoes than any other place on Earth. Oklahoma is in the middle of Tornado Alley, an area of the United States that has frequent twisters. Other countries have twisters, too, but not as often. In fact, Bangladesh, a nation in south-central Asia, had a devastating tornado in 1996.

Looks like our storm is breaking up. That's no surprise. There aren't tornadoes in every storm. In fact, we storm chasers do a lot of driving and study lots of weather data. I usually see only two or three tornadoes in a year.

When a twisting funnel cloud drops down, look out! You have a tornado!

Breakthroughs!

We can always depend on one thing about the weather—it changes. So it's not surprising that people have always observed and recorded weather conditions. And they have always dreamed of predicting or even changing the weather. Hundreds of years ago, scientists began to make the discoveries and inventions that led to modern weather forecasting.

Italian scientist Galileo invented the thermometer. The first thermometer was a glass tube containing air and water. As the air temperature rose, the liquid in the tube expanded and was forced higher. Galileo's thermometer was not very accurate.

1593

1590

Hair hygrometer

1783

A Swiss scientist, Horace Benedict de Saussure, invented the first hair hygrometer. A *hair hygrometer* uses strands of human hair to detect the amount of moisture in the air. Hair expands when the air is moist and contracts when it is dry. Scientists still use hair hygrometers today.

1856-1890

France established the first national weather service, linked by telegraph, in 1856. Great Britain established its weather service in 1860. The United States Congress set up the Weather Bureau (now the National Weather Service) in 1890.

Satellite image of a severe storm over the British Isles

The first successful weather satellite— *Tiros 1*—went into service in 1960. Today, numerous weather satellites constantly photograph Earth's weather patterns from their orbits in space. Satellite photos provide marvelous pictures of large weather systems, such as hurricanes.

1960

NOW

1950's

Meteorologists began using computers to help them forecast weather. In computers, a *modeling program* predicts future weather based on patterns in vast amounts of data from the recent past.

1940's

Scientists began using radar in weather forecasting. Radar detects precipitation, such as rain and hail, in a cloud. It is especially useful for tracking intense storms.

Mathematician John Von Neumann and an early computer for weather research

135

Weather Bulletins from the Past

I'm Sherlock Storms, climate detective—or *climatologist,* if you want to get technical. It's my job to figure out what the weather was like on Earth hundreds, thousands, even millions of years ago. Actually, it's not as difficult as it sounds. There are plenty of clues.

Finding out what the weather was like a few hundred years ago is fairly easy. Old weather diaries and almanacs offer terrific data. Believe me, people have always loved the topic of weather.

It's harder to find written weather records if I want to go back a thousand years or more. That's when I might go to California to check out the giant sequoias. These trees are among the oldest and largest living things on Earth. They can live for up to 3,000 years.

Rings across the trunk of a giant sequoia reveal what the weather was like for every year of the tree's life. A wide ring means that the weather was warm or moist and the tree grew well that year. A narrow ring means that the weather was cold or dry, so the tree grew little.

A page from a weather journal that was kept more than 200 years ago!

Another way to find out about Earth's climate of the last few thousand years is by digging in Europe's peat bogs. *Peat* is a mass of brownish, decayed plants like moss. In some places, peat collects in bogs that are thousands of feet deep. Peat preserves plant pollen. Suppose I dig 100 feet (30 m) into a peat bog. From certain clues, I know that this layer is about 6,000 years old. As I examine it, I find lots of pollen from shrubs and trees that grow in wet, rainy climates. I can conclude that the climate here was wet 6,000 years ago.

But how do we learn about the weather on Earth millions of years ago? Fossils tell us a lot. These remains of plants and animals are found in layers of rock. A fossil reveals something about the weather during the time it was living.

For more information on plants, see chapter 3.

AHA!

Scientists don't have to cut down or kill a tree to examine its annual growth rings. Scientists use an *increment borer*—a long, thin metal tube—to *bore* (drill) a hole in the tree trunk, and then they take out a core sample of the rings.

I don't do all the dirty detective work. This person in Greenland is cutting ice samples.

For example, fossils of palms and ferns tell us the climate was hot and steamy. Fossils of sea animals tell us an ancient sea once covered the area.

Some scientists have even gone to Antarctica (brrr!) to investigate Earth's past. The ice there is very thick—up to 15,700 feet deep (4,800 m). And deep down, it is very ancient. The scientists drill deep holes to find ice that formed millions of years ago. Air trapped in this ancient ice reveals what Earth's atmosphere was like then. For example, scientists found ancient air that contained much less carbon dioxide than air does today.

When I put all these clues together, a fairly clear picture of Earth's past weather comes into focus. For most of the past 500 million years, Earth's weather has been warm and mild. But the fine-weather times, or *interglacial* (IHN tuhr GLAY shuhl) periods, were interrupted by ice ages, called *glacial* (GLAY shuhl) periods. In glacial periods, thousands of feet of ice covered much of the land.

Workers at the National Core Laboratory in Boulder, Colorado, store ice for further testing. Even a snowball would keep in this room!

Of course, we don't live in a glacial period now. But it's important to study the slow changes in Earth's weather. These patterns can reveal future conditions. Will we have another glacial period someday? Will the greenhouse effect warm the earth up? All I can say is: "Stay tuned."

Weather Folklore

Good morning! I hope you're not planning to go sailing. Just look at that red sky. It reminds me of the old saying: "Red sky at night, sailors' delight. Red sky in morning, sailors take warning." You say you never heard that before? Well, I collect weather proverbs. *Proverbs* are sayings that generations pass down by word of mouth. Some proverbs are based on scientific ideas, but others, like this one, are not. Yet people still like to quote them. Let me share some more with you.

"Hen scratches and filly tails:
Get ready to lower your topsails."

Scientific basis? Yes. "Hen scratches" and "filly tails" are folk terms for cirrus clouds. *Cirrus* are very high clouds made of ice crystals. They travel fast and usually warn of rain approaching within 24 hours. Sailors watch for these clouds.

"A bad winter is betide (will happen)
If hair grows thick on a bear's hide."

Scientific basis? No. Animals grow heavy coats when they have lots to eat in summer. Growth of fur has nothing to do with the coming winter.

"Year of snow,
Crops will grow."

Scientific basis? Yes. A heavy snowfall protects plant roots in the ground from frigid air. Also, snow that stays on the ground in spring keeps tender blossoms from opening too early.

"When sheep collect and huddle,
Tomorrow we'll have a puddle."

Scientific basis? Yes. Animals sense the changing weather and become nervous. They huddle together for comfort.

"Seaweed dry, sunny sky;
Seaweed wet, rain you'll get."

Scientific basis? Yes. A piece of dried seaweed hanging on your porch absorbs moisture from the air. If it turns wet, the air is full of water vapor. Rain may be near.

"If bees to distance wing their flight,
Days are warm and skies are bright.
But when their flight ends near
  their home,
Stormy weather's sure to come."

Scientific basis? Yes. When humidity increases before a rainstorm, bees return to the hive for shelter. In fact, bees usually produce less honey during very cloudy, rainy summers.

"If the crescent moon lies on her back,
She sucks the wet into her lap."

Scientific basis? No. The proverb suggests that no rain will fall when the crescent moon looks like an upright cup, because the moon won't spill water. This idea has no basis in science.

SCIENCE
IN THE
WATER

When you think about our planet, do you picture solid ground made up of rock, sand, or soil? Many people do. But the truth about our Earth is that it is indeed a watery world. More than 70 percent of its surface is covered by oceans. We humans live mostly on dry land, so we think of Earth as the "Land Planet." But water is present even on land and underground.

Learning from Fish

No one has ever built a ship that functions in water as well as a fish. You'd almost expect me to say that, wouldn't you? After all, I'm an ichthyologist (IHK thee AHL uh jist). That's a real tongue twister, but the word just means that I'm a scientist who studies fish.

Fish have been navigating the oceans a lot longer than humans have. And fish are very good at what they do. For example, did you know the following:

Ichthyology (IHK thee AHL uh jee) *comes from a Greek word that means "fish." Ichthyology is the branch of zoology that deals with fish.*

- A fish can turn in a space less than a third of its body length. A ship needs one to three times its length to make a turn!

- Many fish can reverse direction without slowing down, something no ship can do.

How do we know these fascinating facts? We know because of people like Michael and George Triantafyllou, ocean engineers who study how fish swim. They believe that studying fish can help people design better submarines—subs that might be used for things such as deep-sea research or work on offshore oil rigs.

A team led by these engineers has built a robot called "Robotuna" that mimics, or copies,

The robot Robotuna in its tank

the way a real bluefin tuna swims. They used the bluefin tuna as a model because it can swim so fast—about 40 miles per hour (64 kph). Like other fish, the bluefin tuna moves through the water by swinging its long, muscular tail from side to side. The first swing creates a large *vortex* (VAWR tehks), or whirlpool. The second swing creates a vortex that rotates in the opposite direction from the first one. When these two vortexes meet, they create a strong stream of water that pushes the fish forward.

Biomimesis
(*BY oh mih MEE sihs*) *is a new science that involves learning from animals. It comes from two Greek words*—bio, *meaning "life," and* mimesis, *meaning "imitation."*

A propeller on a submarine does the same thing, but not as efficiently as a fish's tail.

The tail also works as the fish's *rudder* (RUHD uhr). A rudder is the part of a boat that makes it turn. Actually, the fish's entire body moves to help it turn. A boat has a rudder for steering, but the *hull* (body) of the boat can't move the way a fish's body can. So the boat doesn't steer as efficiently and smoothly as the fish.

Ocean engineers are just beginning to learn how a fish's body moves. Most fish have two *pectoral* (PEHK tuhr uhl) *fins,* one on each side of the front of the body. The pectoral fins control the diving rate of the fish, allowing it to go up and down in the water as it swims forward. These fins also control how the fish rolls from side to side, and they help the fish prepare for a turn. Submarines have similar structures called *diving planes.*

Ocean engineers have also discovered that fish use their pectoral fins and *ventral* (VEHN

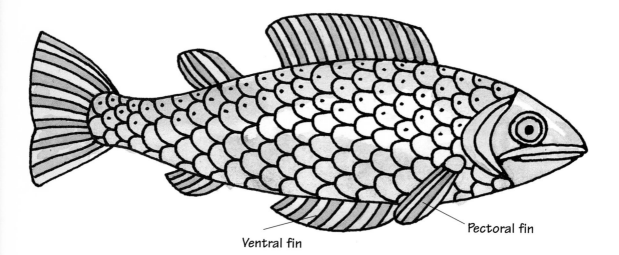

Pectoral fin

Ventral fin

truhl) *fins* for *hovering*—staying in one place in the water. *Ventral* means "belly," so it's no surprise that the ventral fins hang down from the fish's belly. By wiggling their pectoral and ventral fins ever so slightly, fish can make small movements backward or sideways when they are hovering. No one knows exactly how they do this.

When a fish or a ship moves through water, it meets an opposing force called *resistance*. A poorly designed ship needs a lot of energy to overcome resistance and begin to move. But a fish uses very little energy because its body is streamlined. A *streamlined* shape is rounded at the front and pointed behind. You may have seen streamlined bicycle helmets and cars. Submarines and other ships are also streamlined like fish.

When fish are swimming straight ahead, they do something that makes them even more streamlined. They fold their fins against their bodies. This action reduces *drag*—resistance caused by friction from the liquid around it.

Why does it take so much energy to paddle this raft? Because the raft isn't streamlined like a fish. The raft does not have a rounded front and a pointed behind to help it glide easily through the water.

For information on friction, see chapter 6.

Ocean engineers who build small submarines are now thinking about making *retractable* diving planes. That means that the sub's "fins" could be pulled inside to reduce drag when the sub is simply gliding through the water.

As you can see, humans have learned quite a bit from studying fish, and we hope to learn even more.

How a Submarine Floats

Hmmmm . . . I wonder . . . How does a submarine keep from sinking to the bottom? How does it rise to the surface?

GATHER TOOLS

- two plastic 1-liter or 2-liter soda bottles with screw caps
- a plastic straw or a 12-inch (30-cm) length of plastic tubing (You can get this from an aquarium store.)
- a small tub or sink full of water

Set up and give it a try

1 Fill a bottle to the very top with water and cap it tightly. Now put the bottle in the tub of water. What happens?

2 While your bottle is still underwater, take the cap off. Put the straw in the bottle and slowly blow air into it. You will see bubbles rising and a pocket of air in the bottle—your breath is forcing water out of the bottle.

3 Cap the bottle and see what happens. Experiment to get the right amount of air and water inside the bottle so that it will float just beneath the surface of the water.

Try it again and again

- Fill the bottle with hot water.
- Fill the bottle with ice water, but leave out the ice cubes.

Set up a control

Put the cap on an empty bottle. Try to make the bottle stay underwater in your sink or tub. What happens?

Now, write it down

Jot down your observations from your lab. What makes the bottle float or sink? How can you explain this? How is your bottle like a submarine?

Compare your notes with those on page 213.

How a River Changes as It Flows

I'm Faye Delaponde. I'm a geographer—that means I study physical features on the surface of the earth. I specialize in the study of inland bodies of water such as lakes, pond, and rivers. I'd like to introduce you to rivers first. People depend on rivers for drinking water, water for agriculture, and transportation.

The world is filled with important rivers—the Nile in Africa, the Amazon in South America, the Rhine in Europe, the Ganges in India, and many others. Rivers aren't all alike, but they do have some things in common.

Most big rivers start from a small source, such as a mountain *glacier* (mass of ice) or a spring. They follow gravity downhill. And sooner or later, all rivers flow into a body of water such as the sea.

Spring

Tributary

AHA!

The highest waterfall in the world is Angel Falls in Venezuela. It is 3,212 feet (979 m) high. The second highest is Mardalsfossen (Southern) Falls in Norway. It is 2,149 feet (655 m) high.

We'll start our journey near the top of a high mountainside, where a spring spurts out water. The water trickles downhill. Soon several tiny springs join together and form a larger stream. Rainwater adds to the river, too. This water runs downhill, carving out a bed as it goes and picking up speed.

As the rushing stream splashes and foams against rocks, it picks up oxygen from the air. This part of the stream is an oxygen-rich habitat for living things. But they have to swim fast or hold tight! That's why we find strong swimmers such as trout in this mountain stream. We'll also find flatworms, insect larvae, and tiny, plantlike algae (AL jee) clinging to the rocks.

Waterfalls usually occur in mountainous regions, and our growing stream has one. Up ahead the water flows over hard rock down to lower ground. Look at the waterfall plunging down the side of the cliff.

This river isn't done with rough play yet. I see rapids ahead. *Rapids* are a steep, narrow stretch of the river where water rushes very quickly over rocks or boulders. The fast-moving water creates bubbles and sometimes whitecaps.

Now that we're past the rapids, we can go rafting. Hop aboard.

All of the other streams that you see joining the river are called *tributaries* (TRIHB yuh TAYR eez). The river together with all of its tributaries is known as a *river system*. At the same time, *ground water* (water beneath the surface of the earth) seeps into the riverbed. The river grows larger.

Farther along, the land gradually flattens out. The river broadens

AHA!

The tambaqui (tahm BAH kee) is a fish that lives in the Amazon flood plain. It has huge, crushing teeth—more like a horse's than a fish's. These teeth allow it to eat tough fruits and seeds that fall into the water.

Waterfall

Rapids

Meander

and slows down. It *meanders* (mee AN duhrz), or wanders and winds, carving out lazy loops in the level plain. Now we'll get a smoother ride.

Rivers like these sometimes overflow their banks and flood the land nearby. The low-lying land near a river is called a *flood plain*. The flood plain is a food-rich habitat for all sorts of animals and plants, making it ideal for farming.

Our river is not now in flood. But in the past, its waters have risen 10 feet (3 m) above the flood plain. Although a flood like this can mean trouble for people who live nearby, it is good for many living things. Leaves, stems, flowers, fruits, and seeds float around in this watery world. They provide food for animals living in or near the river. They also provide nutrients for plants.

At last, we are nearing the *mouth* (end) of the mighty river, where it flows into a saltwater bay or gulf, or into the ocean. The river begins to widen, and its fresh water mixes with salt water from the sea. This special habitat is called an *estuary* (EHS choo ehr ee). It is a mixed environment—part fresh water and part salt water. The nearby land is low and marshy.

Most animals and plants can live only in fresh water or in salt water. But creatures

AHA!

Can a fish plant a tree? On the Amazon flood plain, about 33 species of trees and shrubs produce seeds or fruits that fish eat. At the end of the fishes' digestive process, the seeds pass out of their bodies. Then they wash ashore and take root. Some of these seeds become great trees of the rain forest.

Flood plain

that live in an estuary can survive in some of both kinds of water. Among these are certain kinds of algae, crabs, shrimp, snails, clams, amphibians, and some fish. Some estuaries nourish an important and unusual ecosystem known as a mangrove swamp. Mangrove trees grow well in the shallow waters of tropical estuaries. Many birds thrive in this environment, as well.

The estuary extends into the river delta (DEHL tuh). A *delta* is a triangle-shaped piece of land that builds up over time at the mouth of a river. Our river has been busy dumping *silt,* extremely small grains of soil, into its delta for many years.

Thanks for exploring this river system with me. We'd better hop out here, or we'll soon be lost at sea.

Mouth

Delta

Estuary

A Model River

Help these students label their science project—a model of a
river system. Pair each label from the pile on the table with the
proper letter label on the model. The answers are listed at the
bottom of the next page.

A. Meanders B. Mouth C. Estuary D. Tributary
E. Delta F. Waterfall G. Flood Plain H. Rapids I. Spring

159

Freshwater
Lakes and Ponds

Hello again! It's Faye Delaponde, and I hear you want to know more about water—specifically fresh water. That's understandable—you and most other creatures living on dry land or in rivers, lakes, and ponds need fresh water. Living things that depend on fresh water would soon die if they tried to consume only salt water. Take a look at these interesting facts:

Glaciers like this one in Antarctica hold three-fourths of the world's fresh water—frozen!

- Of all the water on our planet, about 97 percent is salty, and about 3 percent is fresh.
- Of all the fresh water, three-fourths is locked up in glaciers in Antarctica, Greenland, and a few other frigid places.

For more information on water, see chapter 2.

So, how do we get the fresh water we need? We can take our water out of rivers. We can also drill deep wells to get ground water. In dry places next to the ocean,

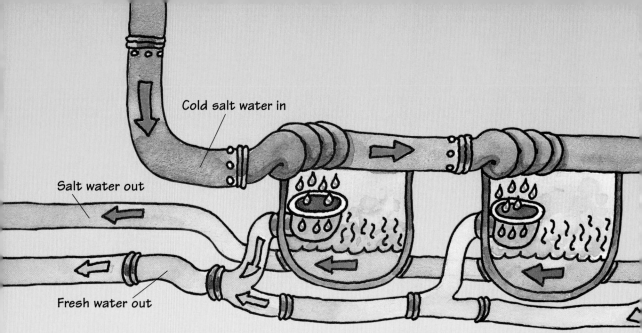

Cold salt water in

Salt water out

Fresh water out

AHA!

Not all bodies of water that we call lakes have fresh water; many actually have salt water. For example, Lake Pontchartrain, in Louisiana, contains *brackish* (partly salty) water. Another example is Lake Eyre in Australia, which is very salty—when it isn't dried up.

Lake Eyre, dry and salty

we can set up desalination (dee SAL uh NAY shuhn) plants. A *desalination plant* is a factory that removes the salt from sea water so that we can use the water.

Lakes and ponds are a valuable source of fresh water, too. Unlike rivers, the fresh water in lakes and ponds doesn't flow—it is surrounded by land and mostly stays put. However, some ponds and lakes dry up because people take out more water than can be replaced. Still other lakes and ponds dry out and refill on a regular basis. Frogs, turtles, and insects can easily migrate to wetter places, but most fish cannot— so they die. To prepare for a dry spell, some fish lay their eggs in the mud. The eggs survive and hatch when rain returns.

All animal bodies contain salt. However, freshwater creatures are saltier than the water

Steam

Heater

they live in. This makes them absorb water. But freshwater creatures have ways to keep out excess water or get rid of it, so they don't swell up and explode. For example, some fish have a waterproof covering of mucus. And some creatures constantly pump out extra water, either through their gills or through urination.

People don't live in water, so we don't have to worry about swelling up and exploding—thank goodness! But we do need lots of fresh water. Our bodies are about 65 percent water, and we need a continual supply of fresh water to maintain vital processes, such as perspiring and eliminating wastes.

Maybe you already knew that water is the most common substance on Earth. Now you know how important—and somewhat rare— fresh water is, too!

One simple way to desalinate water is to boil it and pipe it into a chamber where it turns into steam. The steam rises, leaving the salt behind. As the steam cools, it turns into water again—fresh water.

Fresh Water and Salt Water

Hmmmm . . . I wonder . . . What happens to a freshwater plant when it's put in salt water?

GATHER TOOLS

- 2 or more small, identical goldfish bowls, or similar containers
- 2 or more healthy plants for freshwater aquariums (You can buy these at a pet shop that sells equipment for aquariums.)
- aquarium gravel
- table salt
- a tablespoon measure

Set up and give it a try

1 Cover the bottom of a goldfish bowl with gravel and fill it with water. Plant a freshwater aquarium plant in the bowl.

2 Put the bowl in a sunny place. A window sill or a spot under fluorescent lights will work well.

164

3 Put 1 tablespoon (15 ml) of salt into the bowl and stir gently. Observe the plant each day for a week. Write down what you see.

Set up a control

Set up another bowl with exactly the same amount of gravel and water. Add a freshwater plant but no salt. Label that bowl "NO SALT."

Try it again and again

- Use more or less salt.
- Reverse the experiment. Buy saltwater aquarium plants. Put fresh water in one bowl. Make a salt solution in the other bowl according to the plant instructions.

Now, write it down

Describe or draw what your experiment and control plants looked like at the beginning of the week. Did they look the same or different at the end? Why do you think that was? Can you do another experiment to retest your conclusion?

Compare your notes with those on page 213.

Counting Creatures

If you want to know about the water quality of a river or stream, just ask an insect. Well, you can't *really* ask an insect, but you can find out how clean the water is by studying insects and other creatures affected by pollution.

I'm Flo Brooks, and I'm an environmental scientist. Cleansing pollution from rivers and streams is my specialty. I work with environmental groups, governments, and communities. Together, we decide whether a river is in trouble. If the river is polluted, we work together to clean it up.

I train volunteers to help me. The volunteers check water quality by collecting insects and other small animals from rivers and streams. The mix of creatures tells us a great deal about the health of a stream. In scientific terms,

*An **environmental scientist** studies the quality of land, air, and water as environments for living things. Environmental scientists are especially concerned with the effects of pollution.*

For information on insects in fields and forests, see chapter 3.

we do *biological monitoring* to test water quality. That means we use living things rather than tests with chemicals to judge how healthy or polluted a body of water is.

How can we get so much information from these small animals? Please let me explain.

Many different kinds of insects lay their eggs in fresh water, and the larvae develop there. The larvae of the midge—a kind of small fly—and some other insects can stand more pollution than others. These are called *pollution-tolerant* animals. Larvae of other insects, such as stone flies, riffle beetles, and mayflies, are very sensitive to pollution. Pollution kills them. We call these insects *pollution-sensitive* creatures.

If we find a large number of pollution-sensitive insects in a stream or river, then we know the water quality must be good. But if we find only pollution-tolerant insects, that means trouble! Something has polluted the water and killed the pollution-sensitive critters.

Some animals, such as crayfish, dragonfly nymphs, and clams, are middle-of-the-roaders. They can live in water of just fair quality.

Students stir the stream bed to gather specimens from the bottom of the stream.

168

OK! Now that you understand the basic principles, I'll explain how we collect samples. Volunteers work in parts of a river that are shallow enough to wade in. Of course, we make sure they're properly dressed for this wet work. And every child is teamed up with an adult.

The volunteers use nets to collect specimens—insects or other living things—from the water. One type of net is called a *dip net*. It's just a net on the end of a long handle. It catches anything floating around in the water.

We use a *kick-seine* (kihk-sayn) net to collect samples from the stream bottom. The kick-seine is a mesh net—about 3 feet by 3 feet (1 m by 1 m)—strung between two poles. The volunteer kicks up material from the stream bottom, and it settles in the net. This kicking disturbs more creatures than we catch, but they quickly sink back down and carry on their activities. We also pick up rocks and rub them to collect creatures that cling with claws or suckers.

After collecting specimens in a net, students sort them.

169

The animals we get from the stream bottom help us check the river's oxygen content. Bottom-dwelling animals depend on oxygen that is dissolved in the water. If a good number of these creatures show up in our nets, we know the oxygen content is good. But if we collect very few bottom-living creatures, the oxygen content is probably poor.

The volunteers also identify insects. They carry a magnifying box, a plastic ice tray for sorting, and a "bug card" that has drawings of various river creatures.

Young volunteers play an important part in river monitoring. They have very good eyesight and excellent observation skills. Young children can learn to match a specimen with a picture on the bug card—even if they can't pronounce the critter's name.

Our volunteers do this collecting four to six times a year. And we always go back to the same places, or *testing stations*. Can you guess why?

You're probably wondering what happens at the end of all this testing. Well, environmental scientists like me study the data and draw conclusions about our subject— the river or stream. Then we make recommendations to the communities or governments concerned about water pollution.

A student uses a bug card to identify specimens.

This type of "bug testing" for water pollution works well all over the world. The same kinds of animals live in streams and rivers everywhere. For example, there are more than 400 species of mayflies worldwide, and you'll find at least one or two of them in any stream.

Oops, I'd better go! I've got some more bugs to count. See you soon. Maybe some day you'll be one of my volunteers.

Filtering Water

Hmmmm . . . I wonder . . .
What turns dirty water into
clean water?
How does a filter work?

Set up and give it a try

GATHER TOOLS

- scissors
- absorbent cotton
- clean, small pebbles
- clean gravel
- clean sand
- water
- a small pitcher, such as a measuring pitcher
- a small cup or jar of soil
- two plastic bottles, each 17 ounces (1/2 liter)
- two jars or glasses

1 Cut off the bottom of one of the plastic bottles with your scissors. (Ask an adult to help you.)

2 Tightly pack cotton into the neck of the bottle. Then turn the bottle upside down and put it in one of the jars. The bottle should stand straight and not touch the bottom of the jar.

3 In the bottle, put a layer of pebbles, then a layer of gravel, and finally sand. Use about 1/4 cup (60 ml) of each.

4 Pour 1 cup (240 ml) of water into the pitcher. Add two teaspoons (10 ml) of soil and stir well to make the water dirty.

Caution! Don't drink this water. Even though it may look clean, it still contains germs that could make you sick. A water-treatment plant adds germ-killing chemicals to the water.

5 Slowly pour dirty water into the bottle. If the filter layers seem to sink or mix together, the cotton may not be packed tightly enough. Pack more into the neck of the bottle. Let the water drain through.

— Sand

— Gravel

— Pebbles

— Cotton

Set up a control
To test the usefulness of a filter material—sand, gravel, or pebbles—make another filter without that material and compare your results.

Try it again and again
• Use thicker or thinner layers of the filter materials.
• Pour the water through the filter more than once.

Now, write it down
Jot down how the water looked just before you poured it through your filter. How did the water look after you filtered it? What did your filter take out of the water?

Compare your notes with those on page 213.

Breakthroughs!

Finding clean, fresh water to drink has always been a challenge for human beings. During the last 200 years, people have learned how to filter and treat water to make it safe from germs. Tomorrow's challenge may be to provide safe, fresh water for people living in space colonies, such as on the moon.

Julius Caesar used solar distillation to provide fresh drinking water for Roman soldiers in Egypt. In solar distillation, the sun evaporates water from a pool containing dirty or salty water. The vapor forms droplets of clean water on a dome of glass, and the droplets are then collected.

48 B.C.

40 B.C.

A.D. 100

Around A.D. 100, a Roman official, Sextus Julius Frontinus, wrote *The Aqueducts of Rome,* a book about Rome's impressive water system. Romans used aqueducts to bring fresh, clean water from mountain springs, streams, and lakes to the city.

Painting of ancient Roman aqueducts

The British government built the first desalination plant at Aden on the Arabian Peninsula. This plant supplied fresh water to British ships on long sea voyages.

1869

Scientists in the United States and Germany ran tests using electron beams to disinfect dirty water. In successful tests, the electron beams killed bacteria and viruses in the water.

1995

NOW

1970's–1980's

Scientists developed ways to produce and recycle water on orbiting space labs. This is particularly important on long space missions, when astronauts need lots of water. For example, water vapor from the astronauts' breath can be recaptured, purified, and reused.

This astronaut is surrounded by 100-pound (45-kg) bags of water to be used in space.

Why Rivers Curve This Way and That

(A Korean folk tale)

In a small town with a big pond there once lived a man named Yi Song-gye. He was a famous archer, but he also liked to fish. Every day he used to fish in the pond.

One day, while he was fishing, Yi Song-gye fell asleep. In a dream, he saw an old man with silvery-white hair.

"I am the owner of this pond," the man said, "but a dragon is trying to steal it from me. If you will kill the dragon, I will make you ruler of a kingdom."

"What a strange dream," Yi Song-gye thought when he woke. But as he watched, two dragons, one black and one yellow, rose from the pond. They were fighting furiously.

Yi Song-gye didn't know what to do. As he watched, the dragons sank again and disappeared.

The next day, when Yi Song-gye went fishing, he fell asleep again. Again the old man appeared in a dream. "Why didn't you kill the dragon?" he asked.

"I didn't know which one to kill," said Yi Song-gye.

"The black one," said the old man.

No sooner did Yi Song-gye wake than the dragons appeared again, lashing their tails and striking with their great claws. Yi Song-gye took his bow and shot the black dragon.

The yellow dragon swam away. The black dragon roared in pain and ran toward the sea, twisting and turning its great body. As it ran, it carved a river, and water began to flow from the pond down the twisting, turning path to the sea.

The dragon plunged into the ocean and died. From then on, rivers began to meander—turn this way and that.

The old man never appeared again, but he kept his promise. Yi Song-gye became the ruler of Korea.

SCIENCE
IN THE
CITY

A trip to the busy center of a city is filled with amazing things. Your car rushes down the highway at astonishing speeds. Enormous buildings tower overhead. And glittering lights make the night sparkle. These things seem magical. But magic has nothing to do with them. Science can explain them all.

Is *Friction* a Laughing Matter?

The Downtown Theater has booked me, Fantastic Fred, for two weeks solid. Audiences love my act. My gags and routines help them appreciate a force they meet every day but don't think about very often—friction (FRIHK shuhn).

Watch me try to slide across the stage in my sneakers. What's holding me back? It's friction. *Friction* is the property that objects have which makes them resist being moved across one another. If I roll a ball along the open ground, eventually it slows down and stops even though no apparent force was applied to it. The ball slows down and stops because of friction. Without friction, people could not walk on the earth. Instead, they would slide. When the

The friction of air on moving things is called *drag*. It strongly affects airplanes, birds, and other flying objects. When engineers design new airplanes, they must consider drag in their calculations.

brakes are applied on a car or a bicycle, it is friction that slows down the wheels.

In many instances, people try to reduce friction by making the surfaces of objects move more easily over one another. Watch what happens when I put on my in-line skates. The skates move much faster than the sneakers. The main reason is that the friction of a rolling object is much less than the friction of a sliding object.

If I step on something oily, I will reduce the friction between my skates and the floor even more. Look! There's not enough friction to keep me moving. I'm not getting anywhere, because my skates have lost their grip on the ground. Too little friction can keep me in one place. I need friction to stay on the move.

Have you ever noticed the tires on a mountain bicycle? They have rough rubber surfaces like the soles of my sneakers. These rough surfaces grip the road when a bicycle speeds ahead or turns corners. Without this grip, the tires would spin in place, and the rider would probably lose control of the bike.

Engineers design automobile tires to use friction, too. For instance, in rainy weather, a film of water on a wet road can reduce friction between the wheels and the road. But the *tread,* or raised pattern on the tires, scatters the water. The tires keep their grip, and the driver stays in control.

People who build roads also help tires keep their grip. They cover roads with a rough surface made of sand, gravel, asphalt, and cement. In both dry and wet conditions, this surface produces friction with tires. In addition, engineers give roads a *camber*—a slight upward curve in the middle—to make water run off to the sides. This also prevents water from breaking the tires' grip on the road.

There are many ways that we use friction for safety and control. However, I like to do silly tricks by getting rid of as much friction as I can. Hope you can come see my "Frictionless Follies." The show is even funnier when you see it in person!

AHA!

Maglev (MAG lehv) trains eliminate a lot of friction because they don't have wheels. Instead, they have electromagnets under the cars and on the track. Electromagnets are formed when electric current flows through a wire or other conductor. The magnets repel each other so that the train floats along the track. Soon we may have maglev trains that can travel at 300 miles (480 km) per hour!

IN YOUR LAB Friction

Hmmmm . . . I wonder . . .
Can I measure the amount
of friction between two objects?
How can I reduce the friction?

GATHER TOOLS

- a heavy book (a phone book works well)
- string, about 36 inches (1 m) long
- a rubber band
- 4 juice cans, each 6 ounces (150 ml)
- a ruler or tape

Set up and give it a try

1 Place the heavy book on a smooth floor or table. Attach the rubber band to the string and tie the string around the book. Pull the rubber band to move the book. Measure the length of the rubber band when the book first begins to move. This length shows the work needed to overcome the friction.

2 Place the cans on their sides under the book. Pull the rubber band again. Measure the length of the rubber band when the book first moves. Did it take more or less work to move the book?

Try it again and again
- Use a lighter or heavier book.
- Use smaller or larger rollers, such as pens or larger cans.
- Use rubber bands of different sizes.
- Use a book with a smoother or rougher cover.

Set up a control
Remember, when you test a variable, keep all other parts of the set-up the same. For example, to compare the friction caused by a heavy book and a lighter book, be sure to use the same or identical rubber bands and rollers.

Now, write it down
Jot down your results and observations. Does the rubber band stretch more when the book lies flat or when it rolls on cans? Which variables made it harder to pull the book? Which ones reduced the friction and made it easier? Keep testing your conclusions.

Compare your notes with those on page 213.

What Keeps Skyscrapers from *Falling Down?*

*A **structural engineer** designs bridges, tunnels, dams, tall buildings, and other structures that must be strong enough to carry a lot of weight.*

My name is Spike Towers, and I'm a structural engineer. I have learned the science of using materials to design the buildings we call skyscrapers. I think of skyscrapers as the Olympic athletes of a city—sleek, strong, and impressive.

Some of the tallest buildings from around the world are shown at right. Height is measured from the sidewalk level of the main entrance to the roof. It does not include antennas or flag poles.

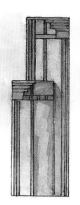

Rialto Tower
Melbourne, 1989
794 feet (242 m)

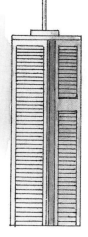

First Canadian Place
Toronto, 1975
952 feet (290 m)

Central Plaza
Hong Kong, 1992
1,227 feet (374 m)

That resemblance is more than skin deep. Like an athlete, a skyscraper has "good bones." Engineers call these bones a *structure*. The part below the ground, which supports the weight of the building, is called the *substructure*. And the part above the ground, called the

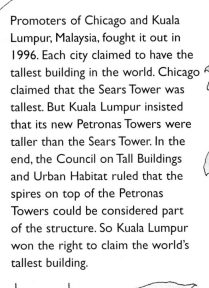

AHA!

Promoters of Chicago and Kuala Lumpur, Malaysia, fought it out in 1996. Each city claimed to have the tallest building in the world. Chicago claimed that the Sears Tower was tallest. But Kuala Lumpur insisted that its new Petronas Towers were taller than the Sears Tower. In the end, the Council on Tall Buildings and Urban Habitat ruled that the spires on top of the Petronas Towers could be considered part of the structure. So Kuala Lumpur won the right to claim the world's tallest building.

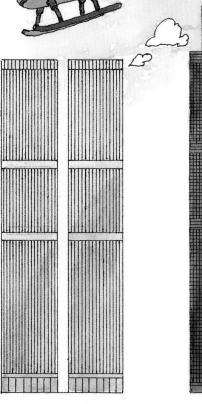

World Trade Center
New York City, 1972
1,368 feet (417 m)

Sears Tower
Chicago, 1974
1,450 feet (442 m)

Petronas Towers
Kuala Lumpur, 1996
1,483 feet (452 m)

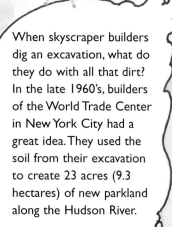

AHA!

When skyscraper builders dig an excavation, what do they do with all that dirt? In the late 1960's, builders of the World Trade Center in New York City had a great idea. They used the soil from their excavation to create 23 acres (9.3 hectares) of new parkland along the Hudson River.

superstructure, gives the building its shape and also supports it.

The substructure begins with an *excavation* (EHKS kuh VAY shuhn)—a hole that may be several stories deep. If solid rock lies at the bottom of the excavation, the builder anchors the structure to the rock. If the ground is soft, the builder drives supporting columns called *piles* down to solid ground and anchors the structure to them.

Before the mid-1880's, the tallest buildings were about 200 feet (61 m) high. It took many iron columns and thick brick or stone walls to support the upper stories of these buildings. The lower stories had only a few windows because glass weakened the supporting walls.

Then in 1885, a steel company developed rolled-steel columns. Finally, builders had a lighter material that was strong enough to hold up a building many stories high. Today, several skyscrapers tower more than 1,000 feet (305 m).

What makes steel such a good structural material? Well, steel stays *rigid* (RIHJ ihd), or stands up, under heavy loads. But it also "gives" a little—that is, it yields to pressure. When weight presses down on steel columns, they shorten without bending. We engineers call this property *strength in compression.*

To demonstrate strength in compression, stand a damp sponge on end. Now press down on it. Notice how the sponge shortens. Steel columns in a skyscraper compress, too, but not as much. You cannot even see the difference.

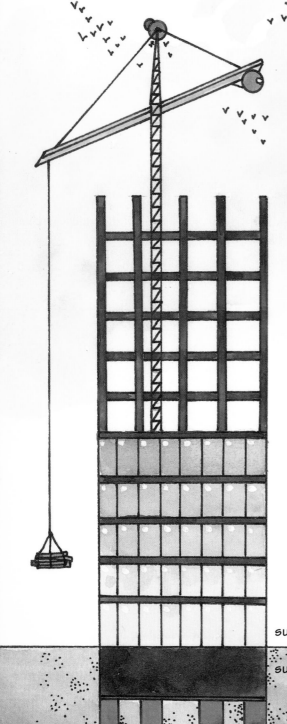

Recently, many skyscrapers have been built with concrete columns. Concrete has strength in compression, too, and it costs less than steel.

Materials used in structures also need to have *strength in tension*. This means that they need to lengthen, or stretch, without breaking—something like a rubber band. Steel is much more rigid than rubber, but it stays strong in tension, too. So builders use steel cables to lift and lower elevators. The weight of an elevator stretches the cable slightly but does not break it.

Builders also use steel to strengthen concrete floors in skyscrapers. By itself, concrete is weak in tension, and it cracks under heavy weight. For this reason, builders pour concrete floors over steel bars or over ribbed steel decks. Steel-reinforced concrete yields

superstructure

substructure

a little under heavy loads, but it does not break. This type of construction is used in roads, too.

But weight is not the only force that acts on structures. Wind pushes on them, too. The force of wind is especially strong on tall buildings, because winds usually blow harder hundreds of feet up than they do at ground level.

Skyscrapers actually sway in high winds. The top of New York City's World Trade Center, for example, sways to and fro by up to 36 inches (91 cm) in a strong wind!

Despite such winds, buildings like the World Trade Center stay upright because their steel structures have the qualities of elastic. Hard to believe, isn't it? Imagine stretching the elastic on a party hat. When you let go, it snaps back. Both steel and concrete are somewhat elastic. When the wind stops pushing, the buildings return to their original positions.

Can you list three qualities that make steel and reinforced concrete "good bones" for a skyscraper? Right! Strength in compression, strength in tension, and elasticity.

Building Up

Hmmmm . . . I wonder . . . How are construction materials used to make a strong building?

GATHER TOOLS

- uncooked spaghetti
- clay, glue, string, tape
- lots of coins
- a hairdrier
- lightweight cardboard

Set up and give it a try

1 Build a box house using the spaghetti and any combination of the clay, tape, and string. Remember, you want your house to be strong in tension and compression.

2 When your house is ready, test its wind strength by carefully blasting the structure from the side with air from the hairdrier. What happens?

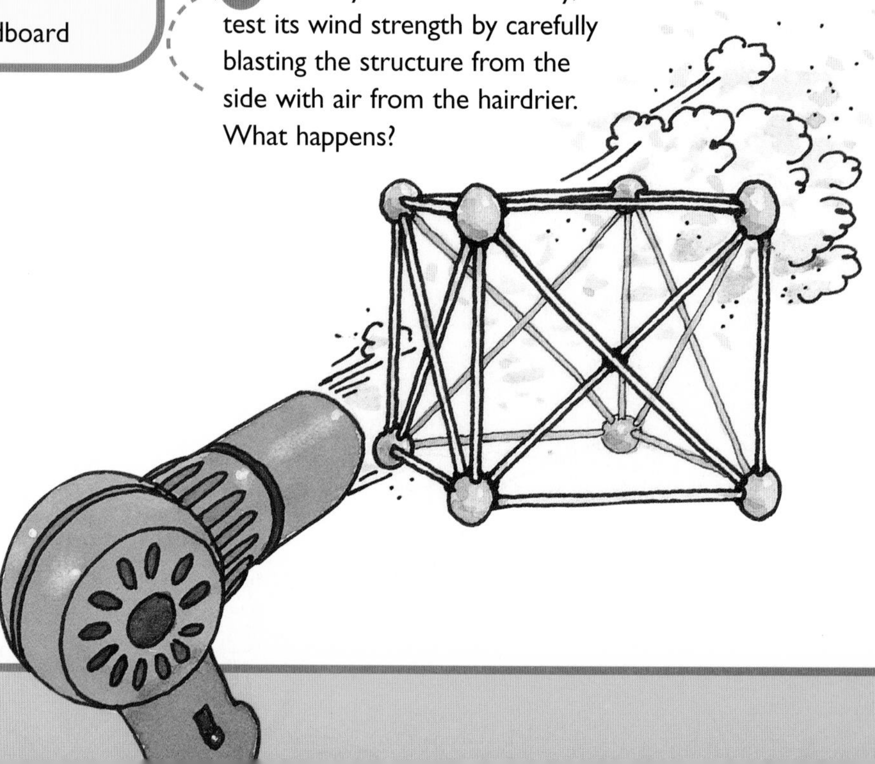

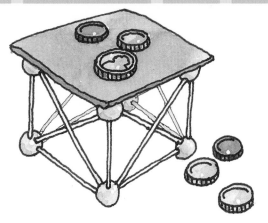

3 Next, cut the cardboard to fit and place it on top of the house. Carefully place coins one at a time on the cardboard. How many coins can the house support before it starts to weaken and then collapse?

4 Rebuild your house to make it stronger. Then test your results.

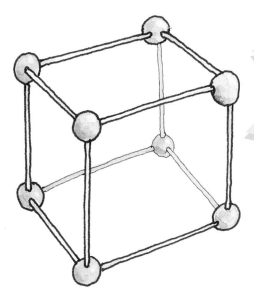

Set up a control
Build a spaghetti house without extra supports— just the 12 outside edges and clay for the joints. Test it.

Try it again and again
• Use more spaghetti.
• Test other materials, such as plastic straws, toothpicks, and empty paper towel rolls.

Now, write it down
What happens to the structures when you pile weights on them? What happens when you blow air on them? Think about the materials and the joints in each one. What effect does each variable have? Jot down your notes and conclusions and keep testing your ideas.

Compare your notes with those on page 213.

A Recycled Building

Hi, Jack Hammer is my name! I'm an architect, but I'm also an environmentalist. Lately, I've been encouraging builders to use recycled materials instead of wood.

I'm doing this because it takes dozens of trees to build a wood frame house. And I'm talking about old trees. Young trees aren't used because they make weak frames. So, forests that have grown for hundreds of years are vanishing.

What do buildings made without wood look like? Let me show you one I designed. We constructed this one-story office building using *rammed earth*. Here's how: You cram soil into a mold and squeeze it hard. The result is a thick, sturdy wall that doesn't require any wood.

The inside walls have steel supports rather than wood ones. The steel came from recycled cars. If more people used steel, we could save millions of trees a year.

For more information on preserving the environment, see chapters 3 and 5.

194

For the inside walls themselves, I used panels made of straw. Unlike particle board, which is made from wood chips, straw panels don't use any tree resins.

There aren't any wood floors here. The lobby floor is made of slate from old chalkboards. In other rooms, I used carpets made from recycled plastic milk jugs!

Get the idea? Look around at the wood in your house. Can you think of recycled materials that might replace it?

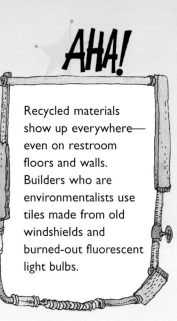

AHA!

Recycled materials show up everywhere—even on restroom floors and walls. Builders who are environmentalists use tiles made from old windshields and burned-out fluorescent light bulbs.

The worker at left is blowing insulation into a wall. The insulation was made by shredding recycled papers, like the ones shown above.

What's Special About a Concert Hall?

Welcome to our city's new concert hall. My name is Melody Acappella, and I study and work in *acoustics* (uh KOOS tihks)—the science of sound. I worked closely with the architect who designed this building to make sure that we'd have a truly great concert hall.

Sound begins when something *vibrates,* or moves to and fro. For example, a violin's strings vibrate when the violinist draws a bow across them. These vibrations create *sound waves* in the air—a bit like the waves a pebble sets in motion when it plops into water.

Sound waves travel out in all directions. Some go directly to listeners' ears. Others *reverberate* (rih VUHR buh rayt)—they bounce off the ceiling, the walls, and the floor. They reach the listeners as echoes.

Rooms in which listeners never hear an echo could be described as "dead." Rooms in which reverberations last a second or more are "live." Music lovers prefer live rooms because they make music sound richer. However, halls with too much reverberation can jumble up the music horribly.

I experienced this muddled sound myself once. I went to a concert where the musicians were playing a fast, lively piece by Mozart. The hall was too "live," so every note reverberated for many seconds. The notes and their echoes jostled and overlapped so much that the beautiful music became a confusing noise.

AHA!

Unlike popular tunes, which generally have many notes and a fast beat, Gregorian chant draws out one note for a long time, without rhythm. Why so slow? Written hundreds of years ago, Gregorian chant was first sung in the great stone churches of Europe. The hard walls of those churches produce reverberation times of up to 10 seconds. Under these conditions, faster music would sound too garbled to enjoy.

That concert hall had many design flaws, starting with flat walls, floor, and ceiling. Its hard, smooth surfaces soaked up very little sound, so they produced long reverberations. And the room had an even number of parallel walls, which didn't allow the sound to spread well to everyone. Why? Because parallel walls tend to bounce sound waves back and forth between themselves. By *parallel* I mean that the walls are the same distance apart their whole length, like the sides of a rectangle.

If too many sound waves reach the listener too loudly or at the wrong time, the music becomes garbled.

Fabric tubes hanging in this music hall absorb sound, while other surfaces help reflect it.

When planning our new concert hall, we made sure we avoided those mistakes. To start, we paid attention to shapes. We chose an odd number of walls to make sound travel throughout the room. We added rounded box seats to the walls because rounded shapes *diffuse* (dih FYOOZ), or spread out, the sound. And the six-sided glass chandeliers hanging from the ceiling do more than shed a lovely glow. Each crystal drop diffuses high notes. We even figured out how much the heads of people in the audience would diffuse the sound. The orchestra sounds better in a packed house!

Nearly everything affects the sound of a performance in a concert hall, including the shapes of the glass chandeliers, the thickness of the carpeting, and the number of people attending.

We also chose surfaces that would add to the music and get rid of unwanted noise. The stage walls were painted with a hard, glossy finish, and the stage floor was varnished to look like glass. These surfaces reverberate the music toward the audience—loud and clear. The rough walls in the rest of the hall diffuse the music. Soft surfaces soak up sound, so we covered the seats in plush velvet to absorb the noise that the audience makes. On the other hand, we did not want too much softness to dim the music. So we covered the floor in a short-pile carpet. It absorbs less sound than a deeper carpet would.

Our concern for sound didn't stop inside the hall. Heavy doors keep out the noise from the

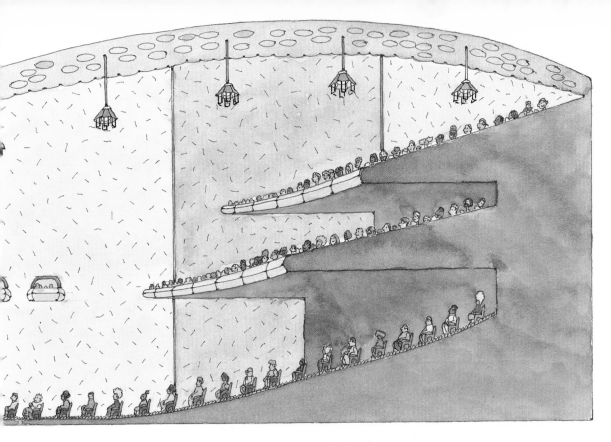

lobby. The lobby itself has walls that help keep out noises such as traffic, horns, and sirens.

So much planning went into this concert hall, and yet it's surprisingly hard to hear me in it. That's because this hall has a reverberation time of two seconds—perfect for musical performances. But people speak much more quickly than musicians play, so the long reverberations cause annoying echoes. In a good room for speeches and plays, a speaker's voice and its reverberations reach the listener's ear no more than one-twentieth of a second apart. This short delay adds sound without jumbling it. Does your school have an auditorium? How was it designed?

Carnegie Hall and other music centers built in the 1800's have rectangular shapes that could produce ruinous reverberations. But true to Victorian fashion, their walls and ceilings have sound-diffusing decorations—busts, plaster angels, fancy woodwork, and murals with lots of paint.

How Bright Are City Lights?

An **electrical engineer** develops electric equipment and machinery.

Let there be light! That's easier said than done—at least, for me it is. I'm Edie Suhn, an electrical engineer, and I've spent my career helping to light up the city.

Many people work to keep the city bright. Architects draw designs that let sunlight shine between buildings. Glassmakers make windows that bring daylight indoors. But when the sun goes down, people everywhere depend on artificial light.

You can create artificial light in two ways. You can make a substance so hot that it glows. Or you can pass an electric current through certain kinds of gas or vapor. Both methods cause electrons (ih LEHK trahnz) to give off light.

Electrons are tiny charged particles that orbit the *nucleus* (NOO klee uhs), or center, of an atom. Heat and electricity make electrons jump into larger orbits. And when the electrons fall back into their original orbits or some other orbit, they give off light energy.

My job is to recommend the best kind of artificial lighting for every purpose. For homes, I usually suggest *incandescent* (IHN kuhn DEHS uhnt) bulbs, like the ones that you probably use. These light bulbs contain tiny coils of *tungsten* (TUHNG stuhn)—a threadlike metal that does not melt at high temperatures. Electricity heats up the coils until they glow white-hot. The coils would soon burn out if they combined with

Giving off light by means of heat is known as **incandescence.**

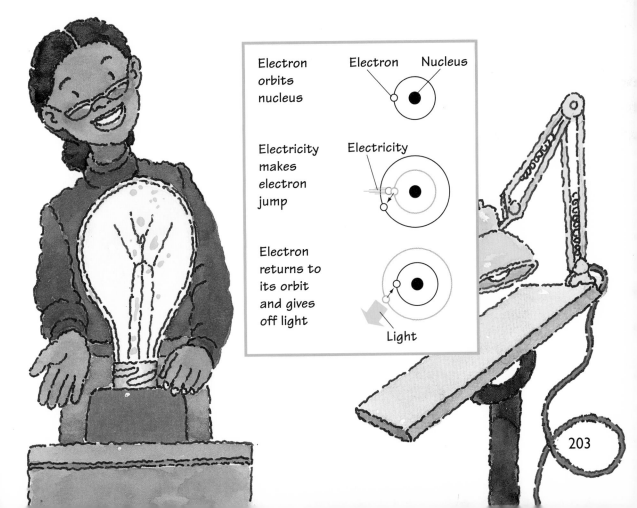

Electron orbits nucleus

Electron   Nucleus

Electricity makes electron jump

Electricity

Electron returns to its orbit and gives off light

Light

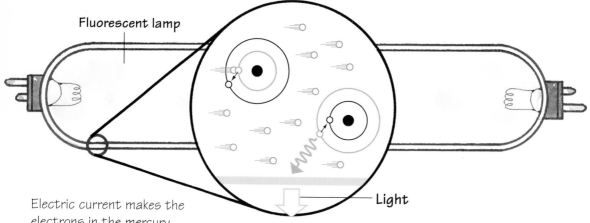

Fluorescent lamp

Light

Electric current makes the electrons in the mercury atoms jump and give off rays of energy. When the rays strike the coated fluorescent tube, it glows.

AHA!

How do certain toys glow in the dark? They are made or coated with *phosphorescent* (FAHS fuh REHS uhnt) substances. Those substances can absorb energy and then slowly give it off again in the form of light for seconds or even days after the energy source has been removed.

oxygen. But an incandescent bulb contains argon or a similar gas that keeps oxygen out—and the bulb stays lit.

People at offices and factories depend on *fluorescent* (FLOO uh REHS uhnt) *lamps*. Here's how they work: electric current travels through a mixture of mercury vapor and another gas inside a glass tube. The electric current makes the electrons in the mercury atoms jump about. As these electrons return to their original orbit, they give off invisible rays called *ultraviolet light*. When these rays strike a special coating on the inside of the tube, the coating glows white.

In offices and factories, where lights stay on for many hours, fluorescent lamps are a wise choice. They use about one-fifth as much electricity as incandescent bulbs to produce the same amount of light.

Street lamps are similar to indoor fluorescent lamps. But most street lamps contain sodium

vapor rather than mercury vapor.
These lamps are even more energy
efficient than mercury vapor lamps.
However, their harsh glare makes
them better suited to outdoor uses.

Street lamps are not the only lights
on the highway. Traffic lights and
auto headlights also help light up the
night. Headlights, like flashlights,
have bowl-shaped mirrors that focus
their light rays into a bright beam.

Reflectors between highway lanes
help light up the road, too. A reflector

Carefully placed lights and reflectors help keep a bicycle rider safe at dusk.

includes a lens and a curved mirror. When a car's headlights shine on the reflector, the lens focuses the light onto the mirror, which bounces the beam back at the automobile. This reflected light helps drivers see the road. In a similar way, reflectors on a bike let drivers know where the rider is located.

My favorite lights are the neon (NEE ahn) signs that brighten shops and theaters. A neon sign maker heats specially coated glass tubes and bends them into shapes, such as letters. Then the tubes are drained of air and filled with neon, argon, mercury vapor, or a mixture of these gases. Like a fluorescent lamp, a neon sign sends an electric current through the tube, lighting up the gas. Neon-filled tubes glow bright red, orange, or yellow, depending on the tube's coating. A little mercury makes a brilliant blue.

Neon signs use a lot of electricity—about 7,500 volts for every 10 to 12 feet (3.3 to 4 m) of tubing. In contrast, a regular 60-watt bulb in your house only uses 120 volts! However, the gas never burns out—unlike the coil in a light bulb—and the tube stays cool to the touch.

Look at all the ways we can light up the night. From the reading lamp in your bedroom to streetlights and sky-high neon signs, science has almost turned the night into day!

Colorful, curving neon lights can be made into almost any shape you can imagine—even a hamburger!

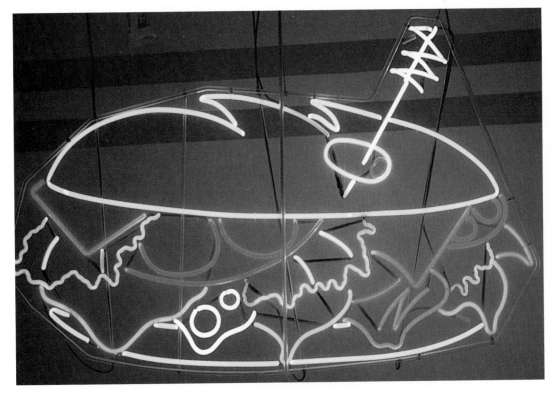

Breakthroughs!

Imagine what the world would be like if you had to depend on the sun, moon, and stars for light. This is what people did in the very distant past. Eventually, prehistoric people learned how to make fire and create light and heat for themselves. Later, they filled hollowed-out stones with melted fat and added a simple wick. Now we have many ways to light up the indoors and outdoors. And we use light for many purposes.

William Murdock, a Scottish engineer, developed artificial indoor lighting from coal gas. Soon gas lamps lit up the homes and streets of towns and cities.

1792

1700

1879

American Thomas Alva Edison invented the first practical electric light bulb. But until electricity became widely available several years later, few people could use light bulbs in their homes.

Thomas Alva Edison

American physicist Theodore H. Maiman constructed the first laser. A *laser* produces a very narrow, powerful beam of light. Lasers are used for many purposes, from performing delicate surgery to playing music on compact discs.

1960's

Corning Glass Works in New York produced the first optical fiber for use in long-range communications. An *optical fiber* is a thin thread of transparent glass or plastic that carries pulses of light energy over long distances. It can carry many more phone calls than a copper wire.

1970

Laser lights on the Empire State Building

NOW

1938

The fluorescent lamp was introduced at the New York World's Fair. Within a few years, it would outsell the incandescent light bulb.

1960's

The light-emitting diode (LED) was invented. An *LED* is a small chip that gives off light when electrical energy excites its atoms. Groups of LED's form letters and numbers in display screens of calculators, digital watches, and laptop computers.

211

Answers & Lab Notes

Answers

Chapter 2 Rhyming riddles

page 39: colloid
page 41: suspension
page 49: surface tension
page 59: gum
page 67: capillarity

page 91 Barry's field guide and drawings

Barry's sketch labeled 1 matches the
dandelion in the field guide. His wildflower
labeled 2 is Queen Anne's lace. Barry's sketch
labeled 4 is a tansy. Don't be fooled by
sketch number 3; it isn't in the field guide!

Lab Notes

Here are some notes and findings you may
have made when doing the labs presented in
this book. There aren't any right or wrong
notes. In fact, you probably made many
observations different from the ones given
here. That's okay. What can you conclude
from them? If a lab didn't turn out the way
you thought it would, that's okay too. Do
you know why it didn't? If not, go back and
find out. After doing a lab, did you come up
with more questions, different from the ones
you had when you started? If you did, good.
Grab your journal and your science kit and
start looking for more answers!

pages 44-45 Mixing Things Up

When a beam of light is shone through
certain suspensions, its path becomes clearly
visible. This occurs because the suspended
particles reflect and scatter light. A solution
shows no such effect because its particles are
too small to scatter light. In a suspension, its
components separate slowly, and they can be
filtered. So, if substances settle overnight,
and solids are found in the filter, then the
mixture is probably a suspension.

Mixtures that do not scatter light, cannot be
filtered, and do not have material that settles
to the bottom are most likely solutions.
Don't be fooled by apparently false readings.
For example, water that is too cold or has
too much sugar in it may appear to be a
suspension because some sugar settles to the
bottom. But it's really a solution. Warm the
water or add more, and then retry the tests.

pages 54-55 Breaking Tension

Because of surface tension, the water acts as
if a thin, elastic film covers its surface.
Water's surface tension is extremely high
compared to many other liquids. Water can
support objects heavier than itself, such as a
carefully placed needle. However, even a
small amount of soap can reduce the water's
surface tension and make the item sink. By
doing the control, you know that it is the
soap, not the toothpick, that reduces the
surface tension.

pages 62-63 Your Own Super-Adhesive

At first, the milk solids do not feel at all
sticky. When the baking soda is added, the
mixture fizzes like soda pop. That is the
baking soda reacting with the vinegar. When
stirred, the solid material turns creamy
white. It makes a good adhesive, especially
on paper and wood.

pages 70-71 Capillarity

Some paper products, such as paper towels,
are designed to soak up and hold water.
However, others, such as coffee filters, are
not. It should be no surprise then that a
coffee filter would soak up water fast, but not
hold a lot of water. Three cheers if you made
this prediction before testing the materials.

pages 80-81 Seeking Light

After several days or weeks, there are distinct
differences in the way the main plant and
the control plants look. In the maze, the
plant bends toward the hole, the light
source. The plant without any light begins to
wilt, turns brown, and eventually dies.

pages 94-95 *Web-Weaver Habitats*

Spiders weave webs in difficult and strange places. Sheet webs and funnel webs are generally found close to the ground. Funnel webs are often built in tall grass or under rocks or logs. Sheet webs are often found between blades of grass or branches of shrubs or trees, even along the edges of streams and lakes. Orb webs, the most complicated webs of all, hang in open areas, often between tree branches or flower stems. Not typically found in shaded areas, orb webs are frequently seen in gardens, meadows, and along roadsides. Tangled webs are often in dark corners, especially in homes. These webs are attached to a support, such as the corner of a ceiling. Dusty, dirty old tangled webs are called cobwebs.

pages 116-117 *Make and Use a Barometer*

When the air pressure is high, the air pushes the balloon down, and the needle moves up. When the air pressure drops, the balloon rises, and the needle moves down. To get the most precise measurements, it helps to put the jars on a stiff board to keep them level.

pages 126-127 *How Warm and Cold Air Behave*

The balloon on the bottle that sits in hot water expands. Nothing is added to the balloon or to the jar, so a person could conclude that the air inside takes up more space when it is heated. In ice water, the balloon contracts and sinks because the cooled air inside the bottle takes up less space. The balloon in the control experiment does not expand or contract, which helps confirm that the change in temperature makes the air—and thus the balloons—take up more or less space.

pages 150-151 *How a Submarine Floats*

The air-filled bottle floats to the surface. The water-filled bottle sinks. By blowing air into the sunken bottle, water is pushed out of it. The underwater level of the bottle can be controlled by adjusting the amount of air and water in it. Submarines control their depth in the water by controlling the amount of air and water in their ballast (BAL uhst) tanks.

pages 164-165 *Fresh Water and Salt Water*

The freshwater plant wilts and, eventually, dies in the salt water. Be sure that the control experiment has the same kind of plant, the same amount of sunlight, and all of the other things the first plant has—except the salt. Then you can be sure that the salt is what makes one plant grow differently from the other. Plants are adapted to living in water of a certain level of saltiness. They don't grow well when there is too much or too little salt in the water.

pages 172-173 *Filtering Water*

Dirty water drips slowly through the layers of the home-made filter. When the water drips out of the bottom layer, it looks clearer and cleaner. The water looks different because the cotton and other materials in the neck of the jar remove some soil and plant material from the water. Purification plants use other materials and chemicals to make water even cleaner—clean enough to drink.

pages 184-185 *Friction*

The more contact there is between the book and the table, the more the rubber band has to be pulled and stretched to move the book. When the juice cans are placed under the book, friction between the book and the table is reduced. The rubber band does not have to be pulled or stretched as much to move the book that is on the cans.

pages 192-193 *Building Up*

Without being properly joined and positioned, even structures of the strongest materials can collapse. Steel and wood are tough, but if they're not joined well or set up in a secure shape, the structure they make is not strong overall. On the other hand, strong joints and a secure shape can make weak materials act like stronger ones. For example, it's easy to snap apart strands of spaghetti. Yet a spaghetti structure can hold extra weight and withstand the force of blowing air if built with strong clay joints and reinforced triangles.

New Words

Here are some words you have read in this book. Some may be new to you. You can see how to say them in the parentheses after the word: **climatologist** (KLY muh TAHL uh jihst). Say the parts in small letters in your normal voice, those in small capital letters a little louder, and those in large capital letters loudest. Following the pronunciation are one or two sentences that tell the word's meaning as it is used in this book.

absorb (ab SAWRB) To soak up or take in.

air mass (AYR mas) A large body of air that has the same levels of temperature and humidity throughout. Air masses travel a long way without changing very much.

architect (AHR kuh tehkt) A person who designs and plans buildings and sees that the plans are followed correctly.

atom (AT uhm) The smallest bit into which a single chemical element, such as gold or oxygen, can be divided without changing it. All substances are made of atoms.

attitude (AT uh tood) A way of thinking or feeling. A scientific attitude is a way of thinking that scientists use in solving problems.

ballast tank (BAL uhst taynk) A tank inside a ship that can be filled with water, air, or other material to balance the ship.

climate (KLY miht) The pattern of weather that is usual for a place. A tropical climate has weather that is usually hot and damp.

climatologist (KLY muh TAHL uh jihst) A person who studies climates and how they change.

compression (kuhm PREHSH uhn) Making something smaller by putting pressure on it or squeezing it.

conclusion (kuhn KLOO zhuhn) A decision reached by thinking and reasoning carefully about all the available information.

control (kuhn TROHL) An experiment that has no changes, or variables, in the materials or in the way the experiment is done. When scientists do experiments with variables, they compare them with the control to see if the results are different.

convert (kuhn VERT) To convert is to change a thing or substance into something else.

214

data (DAY tuh) Information or facts, especially facts from which people can draw conclusions.

dense (dehns) Thick or closely packed.

diffuse (dih FYOOZ) To diffuse is to spread out or scatter something over a wider area.

disperse (dihs PUHRS) In chemistry, to disperse is to scatter bits of one substance evenly throughout another substance.

disposable (dihs POH zuh buhl) Made to be used and then thrown away.

drag (drag) The resistance an object meets when it moves through substances such as water and air.

ecosystem (EE koh SIHS tuhm) All of the plants and animals and all the nonliving things, such as soil, water, and climate, that make up an environment such as a desert, forest, or pond.

elasticity (ih LAS TIHS uh tee) The ease with which a material can return to its original shape after it is stretched or bent.

environmentalist (ehn VY ruhn MEHN tuh lihst) A person who is concerned about problems with air, water, soil, and living things, including problems such as pollution and using up natural resources.

evaporate (ih VAP uh rayt) To evaporate is to change from a liquid to a gas. When water evaporates, it becomes a gas and mixes with the air.

experiment (ehk SPEHR uh muhnt) A test someone makes to see whether a theory or idea might be correct.

filter (FIHL tuhr) A material or device with small openings that is used to clean substances such as air and water. The openings trap pollution or impurities, such as dust or smoke, as the substance passes through them.

fluorescent (FLOO uh REHS uhnt) Glowing by means of ultraviolet light striking a chemical powder on the inside of a bulb. Fluorescent light bulbs give off very little heat.

front (fruhnt) An area where two different kinds of air masses meet.

habitat (HAB uh tat) The place where a plant or animal naturally lives and grows.

humidity (hyoo MIHD uh tee) The amount of moisture in the air.

hypothesis (hy PAHTH uh sihs) An idea that can be used as a possible explanation or answer.

incandescent (IHN kuhn dehs uhnt) Glowing with heat. Incandescent light bulbs contain a thin wire that is hot enough to give off light.

lift (lihft) The upward force created when air pressure under the surface of a moving thing, such as an airplane wing, is greater than the air pressure above it.

meteorologist (MEE tee uh RAHL uh jihst) A scientist who studies Earth's weather and atmosphere.

molecule (MAHL uh kyool) The smallest possible bit of matter that can be made of two or more atoms of elements. A molecule of carbon dioxide has one atom of carbon and two atoms of oxygen.

monitor (MAHN uh tuhr) To keep watch over something.

pollution (puh LOO shuhn) The act of making air, soil, or water dirty by putting harmful substances into it.

precipitation (prih SIHP uh TAY shuhn) Any form of water that falls from clouds, such as snow, rain, sleet, or hail.

property (PRAHP uhr tee) In science, a property is a quality of something. For example, a rubber band has the property of stretching.

resistance (rih ZIHS tuhns) A force that slows down something.

reverberation (rih vuhr buh RAY shuhn) An echoed sound, or sound waves bouncing off something.

specimen (SPEHS uh muhn) A living thing taken from a place as a sample to show what the rest are like.

substance (SUHB stuhns) A kind of stuff or matter.

tension (TEHN shuhn) The condition of being stretched.

tornado (tawr NAY doh) A violent storm that drops down from the clouds to the ground with winds that spin in a narrow, funnel-shaped column.

variable (VAIR ee uh buhl) Something that can be changed, such as the materials used in an experiment or the way the experiment is done, to see if the results will change.

ventilate (VEHN tuh layt) To replace the air in a place with fresh air.

volatile (VAHL uh tuhl) Something that is volatile evaporates rapidly at ordinary temperatures. Volatile oils in spices give them a strong smell or flavor.

Find Out More

There are so many fine books and CD-ROM's about science, you are sure to find plenty to enjoy. The ones listed here are only a sampling. Your school or public library will have many more.

Ages 5-8

Amazing Insects
by Laurence Mound (Alfred A. Knopf, 1993)
Did you know there are about 5 million different kinds of insects? In this book, you will learn how these tiny creatures communicate through sounds and signals. You'll learn about their deadly weapons and more.

The Animals
on CD-ROM for Mac and Windows (Mindscape, 1995)
See and hear more than 200 exotic mammals, birds, and reptiles through video clips as you get information from experts at the San Diego Zoo.

Flowers
by Karen Bryant Mole (Raintree Steck-Vaughn, 1996)
Learn about flowering trees and grasses, flower parts, pollen, seeds and how they spread. This book also tells you how to press flowers.

Monster Bugs
by Lucille Recht Penner (Random House, 1996)
In this "Step Into Reading" book, you'll learn fascinating things about the biggest and fiercest bugs in the world. Did you know that the praying mantis washes its face after it eats?

Sammy's Science House
on CD-ROM for Mac and Windows (Edmark, 1995)
Sammy and his friends will help you develop scientific skills of observation, classification, comparison, sequencing, and problem-solving as you learn about plants, animals, seasons, and the weather.

Songbirds
by Alice K. Flanagan (Childrens Press, 1996)
Did you know that birds use sounds to communicate? About 5,000 species of birds can sing. In this "New True Book," you'll find out why and how this happens.

Water, Water Everywhere
by Mark Rauzon and Cynthia Overbeck Bix (Sierra Club Books for Children, 1994)
Water is part of every living thing. This book explains how, almost like magic, water moves from Earth's surface to the air and back into the ground again.

What's Inside?: Animal Homes
edited by Hilary Hockman (Dorling Kindersley, 1993)
In this book, you will learn all about animal habitats. From the bee's hive to the rabbit's warren, these homes are fascinating!

What's Inside?: Sea Creatures

edited by Hilary Hockman (Dorling Kindersley, 1993)

There's a magical world of life under the sea. You'll read about the shark's internal flotation device, the lobster's soft shell, the blowfish's porcupine look, and much more.

What's It Like to Be a Fish?

by Wendy Pfeffer (HarperCollins, 1996)

This "Let's-Read-and-Find-Out Science" book explains how fish can eat, breathe, and rest underwater. You'll learn how the body of the fish and your body are different and can do different things.

Where Do Puddles Go?

by Fay Robinson (Childrens Press, 1995)

This "Rookie Read-About Science" book explains evaporation, water vapor, condensation, and the water cycle to the very young child.

Why Do Leaves Change Color?

by Betsy Maestro (HarperCollins, 1994)

Learn what chlorophyll and pigment are and their role in how and why leaves turn color. In this book, you'll also learn how to make a leaf rubbing and how to press leaves.

Ages 9 and Up

Adventures with Oslo: World of Water

on CD-ROM for Mac and Windows (Science for Kids, 1995)

This earth science CD explores the role that water plays in supporting life on our planet. The CD includes games, experiments, and animations.

Air: Simple Experiments for Young Scientists

by Larry White (Millbrook Press, 1995)

As you work through the simple experiments in this book, you will learn about air and what it does. You

will learn why warm and cool air make weather, and you will learn how to make a lunch-bag kite.

Into the Sea

by Brenda Guiberson (Henry Holt, 1996)

In this book, beautiful illustrations and text tell the story of a newly hatched sea turtle and take you through her life cycle. The turtle learns survival skills as she encounters many other creatures of the sea.

The Life and Times of the Honeybee

by Charles Micucci (Ticknor and Fields, 1995)

The honeybee is said to be one of the world's most useful insects. This book explains this bee's life cycle, its history, and tells you how it makes honey.

Machines

by David Glover (World Book, 1997)

This book, from the "MAKE it WORK!" science series, is full of simple experiments that will help you explore the workings of all kinds of machines, from bottle openers, pulleys, and gears, to clocks, steam engines, rockets, and much more.

Meat-Eating Plants

by Nathan Aaseng (Enslow Publishers, 1996)

Botanists have identified more than 600 species of meat-eating plants. This fascinating book talks about some of them and tells you how they catch their prey.

Nature in a Nutshell for Kids

by Jean Potter (John Wiley and Sons, 1995)

This book offers more than 100 easy activities to help you discover the wonders of the natural world. These 10-minute-or-less activities are arranged by seasons of the year and cover such topics as cloud formation, leaf decomposition, water speed, and snail study.

Outside and Inside Spiders

by Sandra Markle (Bradbury Press, 1994)

You probably already know that spiders have eight legs and no wings, but did you know that they don't see well? You'll learn why and much more in this book.

Science Fair Projects: The Environment

by Bob Bonnet and Dan Keen (Sterling, 1995)

Do birds prefer popped or unpopped corn? You'll find the answer to this question and much more when you do the sixty projects explained in this book. There is a science glossary at the back of the book.

Sound

by Alexandra Parsons (World Book, 1997)

The simple experiments in this "MAKE it WORK!" book will help you learn about sound, including how your ears work and how sound travels through objects and around corners. You can also experiment with the best way to record sound. Materials needed for the experiments can be found around your home.

Undersea Adventure

on CD-ROM for Mac and Windows (Knowledge Adventures, 1995)

Learn how more than 130 undersea organisms live, what they eat, how fast they swim, and much more with this CD.

Illustration Acknowledgments

The publishers of *Childcraft* gratefully acknowledge the courtesy of illustrator **John Sandford** and the following photographers, agencies, and organizations for the illustrations in this volume. When all the illustrations for a sequence of pages are from a single source, the inclusive page numbers are given. Credits should be read from left to right, top to bottom, on their respective pages. All illustrations are the exclusive property of the publishers of *Childcraft* unless names are marked with an asterisk (*).

Cover:	Aristrocrat, Discovery, and Standard Bindings—John Sandford
	Heritage Binding—Illustrations by John Sandford; Photography by Bob Daemmrich*; Andy Levin, Photo Researchers*
14-15	Bill Bachman*; Andy Sacks, Tony Stone Images*; Bob Daemmrich*
16-17	NASA*; David Gamble, Sygma*
20-21	Corbis-Bettmann*; Michael Nichols, Magnum*
48-49	G. I. Bertnard/Oxford Scientific Films from Animals Animals*
52-53	Archive Photos*; Leonard Ferrante, Tony Stone Images*; University of California, Davis*
64-65	Mel Lindstrom, Tony Stone Images*
68-69	Dan Rest
76-77	Dan Rest
90-91	Larry Mulvehill, Photo Researchers*; Derrick Ditchburn, Visuals Unlimited*; Richard L. Carlton, Photo Researchers*
100-101	D. R. Specker, Animals Animals*
102-103	Thomas Collett*; AP/Wide World*
106-107	Robert Maier, Animals Animals*
112-113	NASA/Science Source from Photo Researchers*
118-119	A & L Sinibaldi, Tony Stone Images*
124-125	Glenn M. Oliver, Visuals Unlimited*; A. D. Copley, Visuals Unlimited*
128-129	National Weather Service*
132-133	A & J Verkaik, The Stock Market*
134-135	Science Museum/Science & Society, London*; European Space Agency/SPL from Photo Researchers*; Princeton University*
136-137	Lancing College Archives*
138-139	Richard B. Alley, Penn State University*; Ken Abbott, Colorado University, Boulder*
144-145	Donna Coveney, MIT News Office*
160-161	Roger Mear, Tony Stone Images*
162-163	Bill Bachman*
168-169	Save Our Streams from the Izaak Walton League*
170-171	Bob Daemmrich*
174-175	Corbis-Bettmann*; High Voltage Environmental Applications, Inc.*; NASA*
194-195	Center for Resourceful Building Technology*
198-199	Arthur Erickson Architectural Corporation*
206-207	Terry Renna; Jon Riley, Tony Stone Images*
208-209	Archive Photos*; Andy Levin, Photo Researchers*; Alfred Pasieka/SPL from Photo Researchers*

Index

This index is an alphabetical list of important topics covered in this book. It will help you find information given in both words and pictures. To help you understand what an entry means, there is sometimes a helping word in parentheses, for example, **aqueducts** (structures). If there is information in both words and pictures, you will see the words *with pictures* in parentheses after the page number. If there is only a picture, you will see the word *picture* in parentheses after the page number.

World Book Encyclopedia, Inc. provides high-quality educational and reference products for the family and school, including a five-volume ***Childcraft favorite set*** with colorful books on favorite topics such as ***dogs*** and ***prehistoric animals***; and ***The World Book/Rush-Presbyterian-St. Luke's Medical Center Medical Encyclopedia***, a 1,072-page, fully illustrated family health reference. For further information, write World Book Encyclopedia, Attention Customer Service, 525 West Monroe, Chicago, IL 60661. Or, contact our website at http://www.worldbook.com